Signals and Communication Technology

More information about this series at http://www.springer.com/series/4748

Bogdan Belean

Application-Specific Hardware Architecture Design with VHDL

Springer

Bogdan Belean
National Institute for Research
and Development of Isotopic
and Molecular Technologies
Cluj-Napoca
Romania

ISSN 1860-4862 ISSN 1860-4870 (electronic)
Signals and Communication Technology
ISBN 978-3-319-65023-4 ISBN 978-3-319-65025-8 (eBook)
https://doi.org/10.1007/978-3-319-65025-8

Library of Congress Control Number: 2017949155

Printed on acid-free paper

This Springer imprint is published by Springer Nature
The registered company is Springer International Publishing AG
The registered company address is: Gewerbestrasse 11, 6330 Cham, Switzerland

To my wife Ramona and to my sons Vlad and Mihnea

Contents

About the Author

Bogdan Belean was born in Tirgu-Mures, Romania, on February 7, 1983. He received the B.E. and Ph.D. degrees in Electronics and Telecommunication Engineering from the Technical University of Cluj-Napoca, Romania, in 2006 and 2010, respectively. Since 2008, he is involved in didactic activates as a research assistant and later as a Lecturer within the Technical University of Cluj-Napoca, Department of Communications. In 2011, Dr. Belean joined the Department of Mass Spectrometry, Chromatography and Applied Physics within the National Institute for Research and Development of Isotopic and Molecular Technology, Cluj-Napoca. In 2014, he became a member of the Center for Research and Advanced Technologies for Alternative Energies (CETATEA) within the same national institute. His research interests include signal and image processing, bioinformatics, application-specific hardware architectures for parallel computing and advanced technologies for alternative energies. His research results include over 30 publications and two open-source software solutions for biomedical image analysis.

Chapter 1
Introduction to Digital Design with VHDL

The first chapter starts with an introduction to digital systems. The levels of abstraction commonly used for the description of a digital system are presented. Each abstraction level includes both a structural description and a behavioral one. The structural description consists of the components and their interconnections used for the digital system design, whereas the behavioral description is used for the representations of the system functionality as a whole. The most important tools in designing digital systems are the hardware description languages (HDLs). They allow the system description at a high abstraction level, where no technology information such as gate level circuit footprints and propagation delays is needed. Meanwhile, based on the HDL descriptions, the synthesis software tools are able to generate more detailed representation of the system, at lower abstraction levels. Considering these abstraction levels, details such as the gate level circuits used by different development technologies (e.g. FPGA or ASIC) are included in the digital system description. Once we established how digital systems are described using different abstraction levels, we proceed to the VHDL language constructs and semantics used to design digital logic. Examples of VHDL codes are provided along this chapter so the reader will get familiar on how to design and test the functionality of digital logic blocks.

1.1 Digital Systems—Introductory Notes

A signal, as referred to in electrical engineering, communications and signal processing, represents a function which gives information about specific phenomena from the physical world. In other words, signals provide information about the variation in time and space of the physical systems. In mathematical terms, signals can be defined as continuous-valued or discrete-valued functions which correspond to the analog or the digital signals, respectively. A digital system is composed of interconnected modules designed to handle digital (discrete) signals in order to

B. Belean, *Application-Specific Hardware Architecture Design with VHDL*,
Signals and Communication Technology,
https://doi.org/10.1007/978-3-319-65025-8_1

analyze and describe specific physical phenomena. These modules are most often based on electronic circuits such as, memories, computing units (e.g. processors), digital audio-video devices or telecommunication devices. The main advantages of such digital modules (devices), where information representation is achieved trough digital signals, are: the reproducibility of information, flexibility, functionality (e.g. easier to store, transmit and manipulate) and the reduced cost. Taking into account these advantages, converting the information into a digital signal (i.e. digitization) has spread to a wide range of applications mainly in the field of computer science, telecommunication and control systems. A major trend in digital design is to use hardware description languages to describe the functionality of digital circuits. The present book is focused in developing digital modules by means of hardware description language, for both real-time image processing application and efficient implementations of channel coders and decoders, specific to the field of digital communication. To understand the methodology for developing digital modules, basic knowledge of digital circuits and their functionality are mandatory. Consequently, simple examples are used in the first chapter in order to ease the reader understanding on the basic concepts of designing digital circuits specific for a given application. Further on, the next chapters present in detail more complex digital circuits, called application specific hardware architectures which fulfill a specific task in a digital system (e.g. real-time edge detection in image processing systems).

A digital system can be analyzed from different perspectives or views: behavioral view, structural view and physical view [1]. The behavioral view examines the system at the most abstract level since it does not take into consideration the internal representation of the system. Practically, it describes the input–output functionality of the system. The structural view specifies how the system is internally represented by its components and their interconnections. This is also known as the diagram of the system. The physical view adds to the system description detailed information like components size, locations on the board or connection line paths. The layout of a printed circuit board is a suggestive example for a physical view of a digital system.

Describing such a complex system using one single process which accounts for all of the systems' views is a complex task. A common approach to ease the design of digital systems is to make use of several levels of simplified models, called abstraction levels. Before proceeding to the description of the abstraction levels, an example is given in order to underline the benefits of this approach. Thus, the process of designing digital logic starts with the behavioral description of the system to be designed. Building blocks such as, registers, multiplexors, and logic blocks with their input output signals are interconnected in order to implement the desired functionality. In this type of description, details such as footprints of the gate level components (e.g. AND logic gate), propagation path delays or other physical characteristics of the target device do not affect the implementation. This description is part of a high abstraction level, where not all the implementation details are accounted. In case of lower abstraction levels, the description of the same digital logic is performed using different building block such as, logic gates and flip-flops, which leads to a more detailed description of the same digital logic.

1.2 Levels of Abstraction

Due to its complexity, a digital system is described by taking into consideration several abstractions levels. Each abstraction level is characterized by (1) the building blocks used to construct the digital system, (2) the representation of the signals that the building blocks operate with, and (3) the behavioural representation of the digital logic functionality.

Considering the building blocks used, the levels of abstraction for digital systems description are [1]:

- Transistor level;
- Gate level;
- Register transfer level (RTL);
- Processor level.

The lowest description level is the transistor level, which includes all the details for the digital systems implementation on the target device and represents the more accurate description. The highest level (i.e. processor level) represents a summarized description of the whole digital system, meaning building blocks such as processors make use of computing units and memories to apply a specific algorithm (computational steps) on the input data.

It is worth mentioning, each digital system can be seen either through its structural description (the building blocks used for its description) or through its behavioural description (behavioural representation of its functionality) [1]. Keeping this in mind, Fig. 1.1 shows both the digital system views and the corresponding abstraction levels.

Considering the transistor level, the main building blocks are transistors, resistors, capacitors etc., whereas the signals are represented by time varying voltages. The behavioural representation and the physical representation are given by differential equations and transistor detailed layouts respectively.

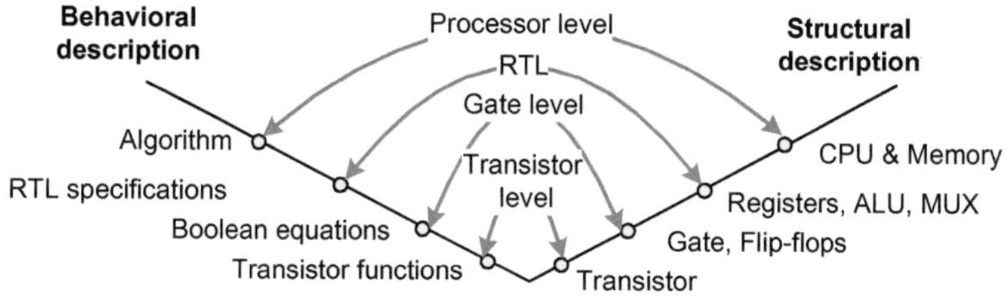

Fig. 1.1 Digital systems views and their corresponding abstraction levels

The second abstraction level, namely the gate level, involves simple logic blocks like logic gates, multiplexers or flip-flop as basic building blocks for digital system description. The signals are represented as logic values 0 or 1, whereas Boolean equations are used for behavioral representation.

The RTL abstraction level has building blocks constructed from simple gates such as arithmetic logic units, multiplexers, comparators or registers. In this case, the signals are interpreted as specific data types, whereas time representation is interpreted using number of clock cycles as time unit. The behavioral representation of the building block from the RTL level is described by finite state machines.

The processor level is characterized by building blocks such as processors, memory modules, intellectual properties and bus interfaces. The behavioral representation at this level of abstraction is performed through a program or algorithm coded in a conventional programming language.

Up to this point, we have a general view on how the digital systems can be described. Thus, the next step is to develop digital systems by using a hardware description language and specific software tools. Developing a digital system supposes the following design tasks to be fulfilled [1]:

(i) Synthesis
(ii) Physical design
(iii) Verification
(iv) Testing

Synthesis

The synthesis process is a transformation of the system either from a description in the behavioural domain to a description of the same design in the structural domain or from a description which makes use of a high-level abstraction to one which uses a low-level abstraction [1]. The synthesis process can be divided into several steps [1], presented as follows: high-level synthesis (i.e. transforms an algorithm into an processor level description with the control and data paths); RTL level synthesis (i.e. transforms the behavioural description of the RTL level to a structural implementation using RTL level components like adders, registers, multiplexers); gate-level (logic) synthesis (i.e. transforms the RTL level description into a descriptions which uses gate-level components and which have the behavioural representations given by Boolean equations); technology mapping (i.e. the gate level circuits are build using the cells of the technology used for the implementation of the digital design under development).

Physical design

The physical design refers to the refinement process between the structural and physical representations, but also an analysis of the circuit's electrical characteristics. The main steps that should be performed are [1]:

- Floor planning—provides a layout at the processor and RT levels; at this step the system is partitioned into function blocks;
- Placement and routing—provides a layout at the gate-level;

Verification

The verification process has to determine if the design accomplishes the specification and the performance required [1].

- Functional verification: check the functionality of the initial design by comparing the output obtained with the output desired; converts the initial design to a gate-level structural representation through the refinement process;
- Performance/Timing verification: the performance is measured by analyzing some timing constraints like maximal propagation delay or minimal clock frequency; at the RT-level the delay of an input-output path is calculated by summing the individual components delays; at the gate-level the propagation delays depend on the components but also on the interconnection wires.

The specific methods of verification for digital designs are:

- Simulation, which examines a system's functionality and performance without effectively building it; practically, the simulation constructs a model of the system but like any model, it involves also limitations (e.g. the simulations does not necessary illustrate the exact same functionality as the real life circuit).
- Timing analysis, which verifies if the system accomplishes the timing goals, by calculating the propagation delays on the circuit paths and by determining the timing parameters;
- Formal verification which is based on mathematical methods for verifying if two representations of a system have the same functionality.
- Hardware emulation, which implements a prototyping circuit which reproduces the system's functionality (e.g. an FPGA circuit may be used to emulate an ASIC design before its prototyping).

Testing

Testing appears to be easy, since the whole design has already been build. Nevertheless, the detection of the physical errors during the fabrication process which affect the functionality of the design is not a trivial task. Thus, considering a large number of inputs and an increased complexity of the designed system, the addition of auxiliary circuits and processes as test pattern generation are demanded.

As a concluded remark, the digital systems are analyzed from different perspectives (behavioural, structural, physical) and they are represented at different abstraction levels. Also, the development of a digital system requires several steps and each step has a very important contribution in constructing a system which meets the specifications and the desired performance. These steps are: synthesis, physical design, verification and testing. Certain tasks of the designing process may be automated, but, the automation process is limited so the digital systems cannot be designed without the amazing work of the human mind.

1.3 The VHDL Hardware Description Language

Knowing that a digital signal can be described from different point of views, perspectives and levels of abstraction we can observe that, the evolution of the design process, is determined both by human minds and software tools. It is a great help to have a standard framework, so that information can be exchanged between peers or software. The common framework is represented by a *hardware description language* (HDL). In the next chapter, a hardware description language will be described in general, so that we get a better picture if its capabilities. The particular case of a hardware description language is VHDL language. A series of language constructs and semantics together with VHDL code examples will be approached to get a closer view of a specific HDL.

1.3.1 Overview of Hardware Description Languages

An HDL is different than any traditional programming language because it's being modeled after hardware. It describes a circuit that was already built or that is in its developing stage. Moreover, with the help of the HDL and synthesis software tools, the circuit can be modeled with high precision at the desired level of abstraction. This means it can be described using any type of buildings blocks (e.g. registers, logic gates or even transistors).

As we already know, there are many programming languages like C, Java, Python and so on, but neither of those are suitable for modeling hardware due to their limitations (they are build exclusively to program general purpose processors). Before the HDL programming languages have been developed, the characteristics of the common programming languages were studied by the designers in order to provide the best syntactic constructs and their associated semantics for building digital logic.

A common programming language is characterized by two features:

- Syntax—syntactic constructs and grammatical rules used to write a program;
- Semantics—the meaning associated with the syntactic constructs.

These two characteristics have definitely been accounted for building the hardware description languages and, naturally, they are found in any HDL. Other than these, there are more differences than similarities.

The majority of programming languages are using a sequential process to describe an algorithm to be used on a classic computer model (for example the C language). In this way the operations are performed one at a time (i.e. sequentially) having two important benefits: at abstract level, it helps the human mind to develop an algorithm one step at a time, whereas at the implementation level, the algorithm being sequential it suitable for the implementation on the basic computer model (one instruction at a given time).

Digital systems work completely different than the sequential ones, having smaller modules with input and output ports, which are connected through customized wiring. These parts/modules with input and output are commonly known as black-boxes. Here we can talk about a propagation delay, due to fact that operations associated with each module are performed concurrently. When signal changes appear at the inputs, new operations are initiated by the black-box for which the inputs changed. When the process is completed, there will be values generated at each output port that can initiate other operations so timing and connection of parts are essential when modeling digital hardware. This is why traditional programming is not suitable for digital systems, leading to the creation of a special language like VHDL, for digital hardware description.

The use of an HDL program

As previously mentioned, HDL programs work in a different way as compared to classic computer architectures. Any hardware description language has three major roles in digital system design:

- *Formal documentation*: at the beginning of the circuit design, a clear description is needed and, because the HDL semantics and syntax are precisely defined, this program (HDL description) will be rigorous and explicit (perfect documentation that can be shared between designers and software tools)
- *Input to a simulator*: before having a physical system, a simulation of the circuit will be needed, making the HDL description ideal to model concurrent operations in a sequential host computer. The input for the HDL description will be generated through a HDL test-bench formed by all test vector generation applied at the inputs of our design described by the HDL program.
- *Input to a synthesizer*: refinement process converts high-level behavioral description to a low-level structural description, those steps being performed by the synthesis software. The input for the synthesis software is the HDL program; the synthesis translates the HDL behavioral description into the structural description using specific libraries for the components used in the HDL program/description.

Before getting into details regarding the particular case of the VHDL hardware description language, we present the common features of a hardware description language in general. Thus the components of a digital logic description will be defined as entities. Connectivity, concurrency and timing characterize each entity and they are also defined in what follows.

The entity is the main independent block modeled after a real life digital circuit which has no information about any other blocks. It describes the inputs and outputs of a given logic block.

Connectivity is the term which defines how different entities interact. More than one entity can be active at the same time, having operations processed in parallel; this is possible because of the way entities communicate with each other through wires; the wires defined by the hardware description language are associated with connectivity parts from real life.

Concurrency characterizes the behavior for all entities placed on a given design. Each entity performs specific operations on the input signals, whereas all operations corresponding to all the entities in the design are executed in a concurrent manner.

Timing specifies the initiation and completion of each operation, providing the order and the schedule of execution in case of multiple operations.

As we have seen so far, digital systems can be described at four different levels of abstraction and, consequently, the hardware description language together with the programming framework (software tools) need to cover all of them. Thus, the language semantics are demanded to encapsulate the concepts of entity, connectivity, concurrency and timing. Moreover, the structural implementation of a circuit can be expressed by language constructs, whereas the operations and structures at the gate and RT levels can be efficiently described by the language. An important characteristic for the description language is to support a hierarchical design process. For example, an entity representing the top level design may be composed of multiple other entities incorporated in the top level one.

The most frequent used HDLs are VHDL and Verilog which have similar capabilities and scopes, even if the syntax of the languages is very different. Both being supported by the same software synthesis tools, they are used in industrial standards, but further on VHDL will be discussed because it is more suitable for parameterized design.

The programming language VHDL stands for VHSIC (very high speed integrated circuit) HDL. It was developed by US Department of Defense as a hardware documentation standard in the 80's, being passed late on to IEEE (Institute of Electrical and Electronics Engineers). After being adopted as a standard, several extensions were developed over the years to meet the latest requirements of the digital design and modeling.

1.3.2 VHDL Code Structure

The VHDL language is constructed from the hardware perspective of digital circuits. The fundamental building block that can be used in a VHDL program is called *design unit*. The skeleton of a basic synthesizable VHDL program consists of three collections of design units:

- library declarations,
- entity declarations and
- architectural bodies associated with the entities.

Each VHDL program is processed according to the following three steps: analysis, elaboration and execution. Consequently, at first the VHDL code is analyzed and translated into design units that are either stored in the libraries or declared as entities. Next step is the elaboration process which designates the top-level entities and associates the corresponding architectural bodies. The last step, execution, creates a single description of the overall design which can be executed, meaning that its functionality can be tested through simulation.

To start with, each type of design units is described, followed by examples which are given to reflect the semantic differences between HDLs and traditional programming languages and to provide the big picture of VHDL.

A *library* represents a collection of pieces of VHDL code which can be re-used or shared between different designs. The pieces of code included in a library are written in the form of components (i.e. entities, data types), functions or procedures grouped inside different packages.

For a library declarations, two identifiers are used, namely "*library*" and "*use*". Note that any VHDL reserved word (i.e. identifier) is mentioned in text using italic fonts (e.g. "*library*"). The example below shows the library definition:

```
library library_name;
use library_name.library_package.package_part;
```

First the library name is specified followed by the package and package parts to be used in the design. Note that, in case of VHDL code sections, VHDL reserved words will be written in bold. The most common examples of library instantiation are *std* library, *work* library and *ieee* library. The first two libraries are included by default in any design and correspond to a resource library and the user defined library, respectively. The first one includes data types and the second one includes all the user defined files. The *ieee* library comprises various packages, out of which the most important are: *std_logic_1164* with standard logic values ('U'—uninitialized, 'X'—strong drive, unknown logic value, '0'—strong drive, logic zero, '1'—strong drive, logic one, 'Z'—high impedance, 'W'—weak drive, unknown logic value, 'L'—weak drive, logic zero, 'H'—weak drive, logic one, '-'—don't care) and *std_logic_arith* which contains the *signed* and *unsigned* data types and related arithmetic and comparison operations together with several data conversion functions.

An entity specifies all inputs and outputs pins of the circuit to be designed. Its corresponding syntax starts with the *entity* identifier and is presented as follows:

```
entity entity_name is
port ( port_name : signal_mode signal_type;
port_name : signal_mode signal_type;
...);
end entity_name;
```

The circuit pins are defined using the *port* identifier. The signal mode can be unidirectional or bidirectional and is defined using the identifiers *in*, *out*, *inout*, or *buffer*. The type of the signal can be for example *bit*, *std_logic*, *std_logc_vector*, etc. More data types will be discussed in detail in the next Sect. 1.3.3, data types and operators.

Finally, the *architecture* for each entity is defined, and it specifies how the designed circuit behaves. The corresponding syntax for any architecture is presented next:

```
architecture architecture_name of entity_name is
[declarations]
begin
(code)
end architecture_name;
```

In the declaration section, the internal signals are defined whereas the code section is composed of concurrent statements and processes. Each concurrent statement or process describes an individual part of the architecture; the architecture can be seen as a collection of interconnected parts of a circuit which are executed in a concurrent manner. If all the architecture statements are concurrent, we cannot say the same thing for the statements within a process. Thus a process is sequential, its statements being executed one after another. More details about processes are presented in Sect. 1.4.2.

1.3.3 Data Types and Operators

We have seen so far that, the entity declaration involves the input/output port description of the design logic circuit, whereas the architecture for each entity describes the behavior of the concurrent statements and processes. Next we will focus on the objects the entities and architectures work with. In this light, three types of objects can be distinguished in case of the VHDL language: *signals*—which represent the interconnection wires to connect the ports of the design unit together, *variables*—which are used for local storage of data, visible inside processes within the behavioral description of the design units and *constants*—which define specific values.

A signal declaration is done as follows:

```
signal signal_name : signal_type [: = initial_value];
```

Signals are declared in entity declaration sections, architecture declarations or in package declarations. The signals are globally visible in all the design entities in case they are declared in the package declaration, whereas if the declaration is done within the architecture section, the signals will be visible only within the architecture they are defined in.

In order to assign a value to signal, the following syntax is used within the architecture of any entity:

```
signal_name <= initial_value;
signal_name <= other_signal_name;
signal_name <= input_port;
```

A declaration and the assignment of a variable look like this:

```
variable variable_name : variable_type[: = value];
variable_name : = value;
```

Variables can be declared only inside architecture or within a sub-program. Note that there is an important difference between variables and signals. The variables are assigned immediately, whereas signals are only scheduled for assignment at the end of the architecture or process. More details about signal assignment scheduling and variable assignment are provided in Sect. 1.3.4.

The three objects discussed so far (the signal, the variable and the constant) can be declared using a type specification. VHDL contains a wide range of types that can be used to be associated with each object. Further on, the fundamental data types of VHDL are presented together with the conversion possibilities between different data types.

Commonly used data types definitions are found in the following packages:

- *standard* package of *std* library with *bit*, *boolean*, *integer*, and *real* data types;
- *std_logic_1164* package of *ieee* library with *std_logic* and *std_ulogic* data types;
- *numeric_std* package of *ieee* library with *signed* and *unsigned* data types.

A type of a VHDL objects is defined by the set of values which may be assigned to the object in question and by the operations that can be performed with these objects. Each of the VHDL objects can be assigned only a value of its type, meaning that VHDL is a strongly typed language.

It is important for the reader to be informed that, not all data types are synthesizable, meaning that digital logic is inferred only to some of the existing data types. A relevant example for an un-synthesizable data type is the *file* data type. An object of type *file* represents a file containing sequential streams of a particular type. A file object can be read from and written to with special procedures and functions. The content of the file may be delivered to the designed logic block using test-benches. The VHDL code for test-benches is not synthesizable and is used only for delivering input and output data to the designed logic blocks. Writing test-benches with VHDL code is discussed in 1.6. Further on we will focus on synthesizable VHDL data types.

Predefined data types of std library

The predefined synthesizable VHDL data types included in the *std* library are:

- *integer* from—$(2^{31}-1)$ to $(2^{31}-1)$ with the subtypes *natural* and *positive;*
- *boolean* defined as false and true;
- *bit:* defined as 0 and 1;
- *bit-vector:* defined as one-dimentional array of the *bit* data type.

The operators associated with the previously mentioned data types are listed in Appendix A.

Standard logic data types

In real life a signal may have also different values than 0 or 1. Consequently, in the *std_logic_1164* package introduces the *std_logic* data type which consists of the following values: 'U', 'X', 'O', 'l', 'Z', 'W', 'L', 'H', '-'. '0' and '1' stand for 0 and 1 logic values. 'U' stands for uninitialized logic value, 'X' and 'W' stand for unknown values, 'Z' stands for high impedance, 'L' and 'H' mean weak 0 logic and weak 1 logic respectively, '-' stands for don't care logic value.

An array of elements with the data type std_logic is called *std_logic_vector.* The dimension of the array is specified in brackets using *to* or *downto* reserved words. Examples of signal declarations having the data type *std_logic* and *std_logic_vectors* are presented next:

```
signal a : std_logic : = '0';
signal b : std_logic_vector (7 downto 0) :="00000000";
```

Considering the VHDL is a strongly typed language, the definition of an object having a specific data types includes also the operators that can be used with the data type in question. Thus, it is important to know the operators that can be used with each data type. For the *std_logic* data types, we can apply any logic operator (e.g. not, and, or, xor, nand, nor, xnor). Regarding arithmetic operators, it is important to know that they cannot be applied. The conversion to an arithmetic data type is needed (e.g. signed or unsigned, discussed in numeric standard package).

If we consider the *std_logic_vector* data type, we have to mention the specific operators for the array data types, namely relational operators, concatenation operators. Examples of relational operators are equal ('='), not equal ('<>'), greater ('<'), smaller ('>'), whereas the concatenation operator is '&', which is used to combine parts or elements of different arrays to form a larger array.

Numeric standard data types

Digital hardware involves arithmetic operations, thus it comes natural to use the integer data types for two *a* and *b* signals considered as the addition terms. Nevertheless, the range of integer not being specified, it is difficult to implement this in hardware. Consequently, *signed* and *unsigned* data types are included n the IEEE *numeric_std* package. These types represent an array of elements having the *std_logic* type. Declaration of such data type is similar with the std_logic_vector, as denoted by the line of code:

```
signal a,b : signed(15 downto 0);
```

The difference between signed and unsigned is that for the signed type, the bits are interpreted as a signed binary number in 2's complement format. These two types support arithmetic operations, whereas the operators to be used are *abs* (absolute value), *, /, +, −, *mod* and *rem*. Moreover, the relational operators such as = , < , > and similar ones can be used with the signed and unsigned data types.

Type conversion

Having the synthesizable data types detailed in the previous sections, the next step is to discuss the conversions between different data types. Type conversion is mandatory in VHDL considering that direct operation between data of different types cannot be performed. Type conversions are performed either using a *type conversion function* or *type casting*. A simple example is given next, in order to have a better view on type conversion. Let's assume we need to access a memory location for which the address is computed by an address computation unit *addr_comp_unit*. In other words, the memory address is given by an arithmetic operation, a multiplication for example. In case the memory address port is of *std_logic_vector* type, the *addr_comp_unit* needs to deliver an output of type *std_logic_vector* for the memory port. The next code example, example 1.1, makes use of type casting in order to convert the multiplication result (i.e. *unsigned* data) into *std_logic_vector* data type.

Example 1.1—Type casting unsigned to standard logic vector

```
library ieee;
use ieee.std_logic_1164.all;
use ieee.numeric_std.all;
entity addr_comp_unit is
        port (
        a : in unsigned (7 downto 0);
        b : in unsigned (7 downto 0);
addr : out std_logic_vector (15 downto 0));
end addr_comp_unit;
architecture behavioral of addr_comp_unit is
signal mult: unsigned(15 downto 0);
begin
        mult <=a*b;
        addr <= std_logic_vector(mult);
end behavioral;
```

Note that for the previous example the *numeric_std* package was used. Considering *numeric_std* package, the type conversions of numeric data types are performed using the conversion functions or type casting operators summarized in Table 1.1.

Table 1.1 Conversion functions for VHDL data types

From data type	To data type	Conversion function/type casting
unsigned, signed	*std_logic_vector*	*std_logic_vector()*
signed, std_logic_vector	*unsigned*	*unsigned()*
unsigned, std_logic_vector	*signed*	*signed()*
Unsigned, signed	*integer*	*to_integer()*
natural	*unsigned*	*to_unsigned()*
integer	*Signed*	*to_signed()*

1.4 Combinational Logic, Sequential Logic and VHDL

Combinational logic refers to circuits whose output is a function of the present value of the inputs only. Whenever the inputs are changed, the information about the previous inputs is lost, meaning that a combinational logic circuit has no memory. It can, therefore, easy be implemented using conventional logic gates.

Sequential logic circuits are those whose outputs are also dependent upon current and previous inputs. This means that they have memory, so storage elements are required, which are connected to a combinational logic block through a feedback loop. In this way, the stored states, created by previous inputs, will affect the output of the circuit. Note that not any circuit that has storage elements is a sequential circuit. For example, memories obviously store data, but their output depends only on the address bits applied as input data. A better view on the two types of logic circuits can be depicted in Fig. 1.2.

Corresponding to the two types of digital logic, combinational or sequential, the VHDL code is classified either as *concurrent* or *sequential* [2]. In what follows both concurrent and sequential VHDL code are detailed and code examples are also provided.

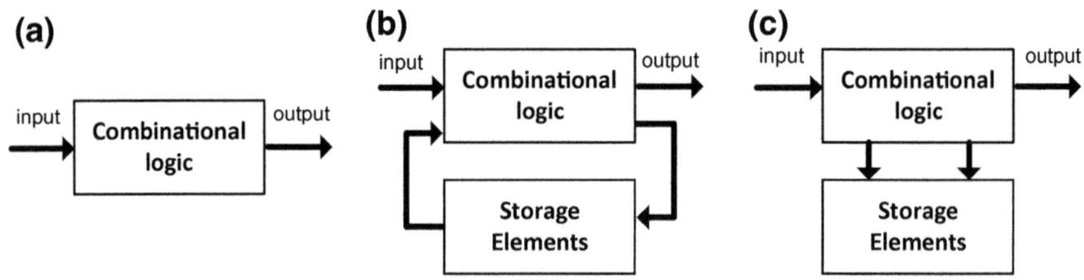

Fig. 1.2 **a** Combinational logic representation, **b** sequential logic representation, **c** memory representation

1.4.1 Concurrent VHDL Code

In general, all VHDL code is concurrent, meaning each line of code placed inside the behavioral description of logic blocks is executed in parallel. The only exceptions are statements placed inside a PROCESS, FUNCTION, or PROCEDURE which rerpesent the sequential VHDL code and which are detailed in the *Sequential VHDL code* section. Within curret section, the concurrent VHDL code know also as *dataflow* code is discussed.

The VHDL statements used to create combinational circuits (i.e. concurrent VHDL code) are: (i) the assignments using different type of operators (logical, arithmetic, etc.), (ii) the WHEN statement and (iii) the GENERATE statement. Example for each type of cocnurrent VHDL code are given next.

(i) Within current paragraf the description of a *Gray code converter* is presented as an example of combinational logic circuit. In 1947, Frank Gray from Bell Laboratories, introduced the term reflected binary code in a patented application, based on the fact that it may be built up from the conventional binary code by a sort of reflexion process. The main feature of this code is that a transition from one state to a consecutive one, involves only one bit change. The conversion procedure from binary natural to Gray is the following: the most significant bit, MSB, from the binary code is the same with the MSB from the Gray code. Starting from the MSB towards the least significant bit, LSB, any bit change (0 to1 or 1 to 0) in binary natural, generates an '1' and any lack of change generates a '0', in Gray code. The conversion from Gray to binary natural is the reverse: the MSB is the same in binary natural code as well as in Gray code; further on, from MSB to LSB, the next bit in binary natural code will be the complement of the previous bit, if the corresponding bit from Gray code is 1 or, it will be identical with the previous bit, if the corresponding bit from Gray code is 0.

In order to build a combinational logic circuit which transforms any 4 bits input vector $[b_3b_2b_1b_0]$ from binary natural representation into Gray code representation $[g_3g_2g_2g_0]$, the truth table for the binary to Gray conversion is build. Based on function minimisation, each binary output g_i is expressed as $g_i = f(b_3, b_2, b_1, b_0)$, where f is a logic function of the 4 b_i inputs. Consequently the equations for a binary to Gray conversion in case of a 4 bits logic vector are as follows:

$$\begin{aligned} g_3 &= b_3 \\ g_2 &= b_2 \oplus b_3 \\ g_1 &= b_1 \oplus b_2 \\ g_0 &= b_0 \oplus b_1 \end{aligned} \tag{1.1}$$

The VHDL code for the combinational logic cirrcuit (CLC), which describes the binary to Gray conversion according to previous equations set, is presented. Firstly, the *CLC* entity describes the input and output ports of the binary to Gray converter.

Thus, the *BN* is a 4 bits input vector, whereas the *Gray* output returns the Gray code representation of the *BN* input. Once the entity is described, we may proceed to the behavioral description of the circuit. In our case, the architecture of our entity includes concurrent VHDL code which includes 4 assignments, each coressponding to one of the output bits. Moreover, the *xor* operator is also used to describe the set of conversion Eq. (1.1).

Example 1.2—Concurrent VHDL code for binary to Gray converter

```
entity CLC is
     Port ( BN : in STD_LOGIC_vector (3 downto 0);
     Gray : out STD_LOGIC_vector (3 downto 0));
end lab1;
architecture Behavioral of CLC is
begin
     Gray(3) <= BN(3);
     Gray(2) <= BN(3) xor BN(2);
     Gray(1) <= BN (2) xor BN(1);
     Gray(0) <= BN (1) xor BN(0);
end Behavioral;
```

The previous concurrent VHDL code corresponds to the combinational logic circuit descibed in Fig. 1.3.

(ii) Another concurrent VHDL code is the *when* statement. Commonly, these statements are used when multiplexers (i.e. combinational logic) need to be described with VHDL. Moreover, the tri-state buffers are also combinational logic circuits that can be described using *when* statements. Examples of both multiplexer and tri-state buffer are given next. Before proceeding to the examples description, it is to be mentioned the two forms of the *when* statemet. There is the simple *when /else* statement and the *with /select /when* statement also known as *selected when.* The syntax for the two when statement is provided next:

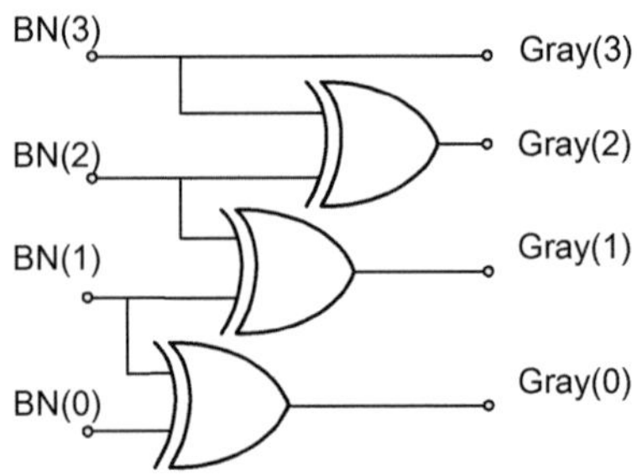

Fig. 1.3 Combinational logic for binary to Gray conversion

- for the simple *when /else* statement

```
assignment when condition else
assignment when condition else
...;
```

- for the selected when statement (*with /select /when*)

```
with identifier select
assignment when value,
assignment when value,
...;
```

The simple when /else statement is used for the descrption of the tri-state buffer presented in Fig. 1.4a, whereas the multiplexer from Fig. 1.4b is described using a selected when statement.

Example 1.3—VHDL code for multiplexer

```
library ieee;
use ieee.std_logic_1164.all;
entity multiplexer is
port ( a, b, c, d: in std_logic;
  sel: in integer range 0 to 3;
  output: out std_logic);
end mux;
architecture mux of multiplexer is
begin
  with sel select
  output  <= a when 0,
    b when 1,
    c when 2,
    d when 3;
end mux;
```

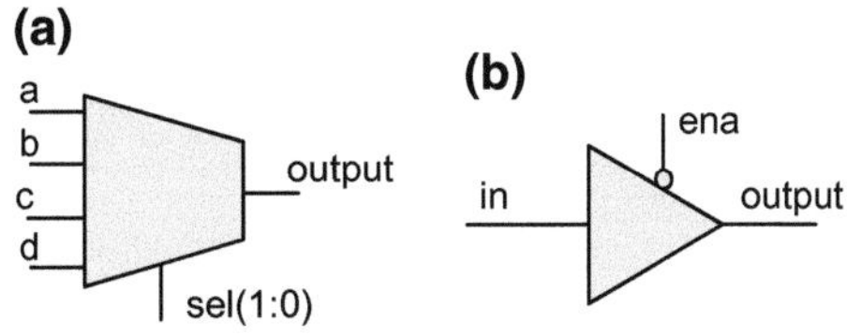

Fig. 1.4 **a** Multiplexer, **b** tri-state buffer

The behaviour of the logic crcuit described by the previous VHDL code is as follows. The *output* is one of the 4 inputs *a, b, c* or *d,* depending on the *sel* input. If the *sel* input is *0*, the output is *a*; if the *sel* input is *1* the *output* is *b* and so on. Note that in case of the *with /select /when* statement all the possibilities for the *sel* input must be tested. This is why the keyword *others* is often used as one of the *sel* identifier values (e.g. **with** sel **select** output <= 0 **when** *others*).

Example 1.4—VHDL code for multiplexer

```
library ieee;
use ieee.std_logic_1164.all;
entity tri_state is
  port ( ena: in std_logic;
    input: in std_logic;
    output: out std_logic);
end tri_state;
architecture tri_state of tri_state is
begin
  output  <= input when (ena = '0') else 'z';
end tri_state;
```

This examples ilustrates another way of using the *when* statement. The tri-states logic circuit provides the *input* at the circuit *output* if the *ena* is '0' and high impedance 'Z' otherwise.

(iii) the *generate* statement is another concurrent statement used for repeating a section of code for a number of times. In this way multiple instances of the same assignment are created. In other words, generate statement may be used to replicate logic. The syntax for this type of concurrent code is specified below, whereas further on, an example (1.4) on how to use generate statement to replicate logc is provided.

```
label: for identifier in range generate
(concurrent assignments)
end generate;
```

Example 1.5—Generate statement to replicate logic

```
library IEEE;
use IEEE.STD_LOGIC_1164.ALL;
entity converter is
  Port ( input  : in   STD_LOGIC_VECTOR (7 downto 0);
      output  : out  STD_LOGIC_VECTOR (7 downto 0));
```

```
end converter;
architecture Behavioral of converter is
begin
 U1: for i in 0 to 7 generate
 begin
    output(i) <=input(7-i);
 end generate;
end Behavioral;
```

The previous VHDL code example describes a combinational logic circuit which takes as input a logic vector *input* and returns another logic vector, the *output*, with the same bit values as the input but in reverse order. The code line *output(i) <=input (7-i)* assigns to the output bit *output(i)* the logic value from *input(7-i)*. The generate statement multiplicates the previous assignments for all *i* values from *0* to *7*, the result beeing the *output* logic vector with the bit values from the *input* in reverse order.

1.4.2 Sequential VHDL Code

As mentioned before, VHDL code is concurrent, all the statements placed in logic blocks architectures being executed in parallel. Nevertheless, it is also important to have statements executed one after another; this is the case of sequential circuits. A common example is a shift register which sequentially changes its content each clock cycle by means of bits shift. Also, arithmetic operations need to be executed one after another since one operation may depend on the results of a previous one (see the logarithm computation unit from Chap. 3). In order to describe sequential logic with VHDL, the statements have to be included within a *process*, *function* or a *procedure* section. The statements placed in one of these three types of code sections are sequential. Notice that not any kind of statement can be included in processes, function or procedures; the statement are restricted to *if*, *wait*, *case*, *loop* together with *variables* and *signals assignments*. Further on, a process is defined in the context of VHDL coding and also, all sequential statements (signal and variable assignments, if, wait, case and loop statements) which can be placed within a process, are defined.

The *process* statement or simply a process is composed of 3 parts: sensitivity list, process declarative part and the statements part. The process begins with the *process* keyword, followed by a parenthesized list of signals called the sensitivity list. The process is activated on any change of the signals in the sensitivity list. After the process sensitivity list, the declarative part comes followed by the sequential statements part. The syntax for a process declaration is presented next.

```
[label:] process (sensitivity list)
variable declarations;
begin
(sequential code)
end process [label];
```

Remark

Answering the question "*How the processes are executed?*" is the key for understanding the VHDL code. We have so far how sequential logic is described using processes. How these processes are executed is explained next. A process execution is triggered by any change in the signals from the sensitivity list. Once a change occurs in the sensitivity list, the process starts. *Variable* and *signals* are assigned during the process execution in case some given conditions are met (e.g. if reset = '1' then a signal or variable is initialized with a given value). Commonly, these variables and signals are passed to the output of the entity from which the process belongs. Notice that there is an important difference concerning variables and signals assignments within a given process. Thus, in case of signal assignments, the assigned values are only *scheduled* for instantiation; the scheduled values are instantiated *only at the end of the process*. On the other hand, in case of variable assignment, the assigned values are immediately instantiated to the variable in question, their new values being available in the next line of code.

Signal and variable assignments

Signal and *variable* assignments are used to pass non-static values between logic components described with VHDL code. Signals can be declared in a package, entity or architecture, whereas variables can be declared only inside a sequential code section (process, function or procedure). Considering the signals assignment, it can be done either inside or outside of a process. In the first case, combinational logic circuits are described, meaning the assigned value is instantly visible. In case signals are assigned in a process, the assigned value is available only after the conclusion of the process run. Notice that for signal assignment the operator is "<=". Variables assignment on the other hand, can only be performed within sequential code section. The update of a variable is immediate, thus the new value is available promptly after the assignment. Another important aspect regarding variables is that they represent local information, and they are only visible inside the process they are assigned in. In order to pass values outside the process, the signals are used which are globally declared within the architecture declaration section. For variable assignment the operator is ":=". Further on two examples are provided in order to underline both the signal and the variable assignment. Both examples are used to describe a counter used to deliver input data to the combinational logic circuit aiming for a binary to Gray conversion (Fig. 1.3). The counter and the combinational logic block for the binary to Gray conversion are both illustrated in Fig. 1.5.

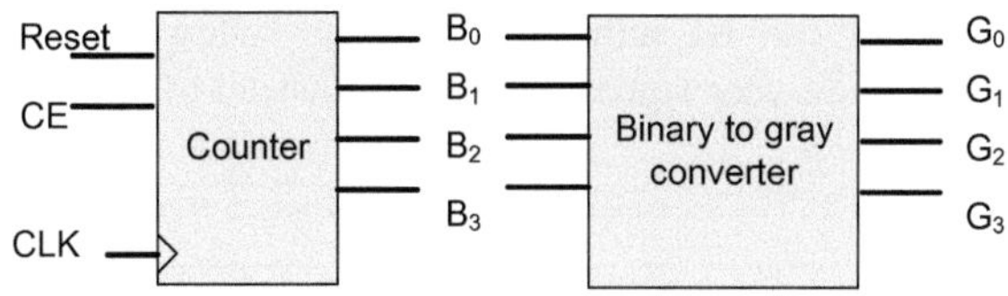

Fig. 1.5 Conversion of 4 bits binary numbers to Gray code representation

Example 1.6—Counter description with variable assignment

```
library IEEE;
use IEEE.STD_LOGIC_1164.ALL;
use IEEE.STD_LOGIC_ARITH.ALL;
use IEEE.STD_LOGIC_UNSIGNED.ALL;
entity cnt is
  port ( clk: in std_logic;
       CE : in std_logic;
        reset : in std_logic;
        Counter: out std_logic_vector( 3 downto 0)) ;
  end cnt;
architecture Behavioral of cnt is
begin
  process (clk, reset, CE)
  variable temp_a: std_logic_vector (3 downto 0) : = "0000";
  begin
    if reset = '1' then
        temp_a: = "0000";
elsif clk = '1' and clk'event then
      if CE = '1' then
        temp_a : = temp_a + 1;
      end if;
    end if;
    counter <= temp_a;
  end process;
end behavioral;
```

The logic circuit name *counter* described by the previous VHDL code is used to consecutive binary numbers to the binary to Gray converter. The inputs are *CE* (count enable), *reset* and *clk,* which are found also in the sensitivity list of the process used to increment or to reset the counter output. The '1' logic value on the *reset* port sets the counter output to "0000", whereas the '1' logic value on *CE* port enables the counting, meaning each *clk* cycle the output value is incremented by 1. The variable *temp_a* is used for incrementing, and its value is passed outside the process to the counter output *counter* by the following assignment *counter <= temp_a*. Similar behavior can be achieved using the following VHDL code, where the counter description uses signal assignment instead of variable assignment.

Example 1.7—Counter description with signal assignment

```
library IEEE;
use IEEE.STD_LOGIC_1164.ALL;
use IEEE.STD_LOGIC_ARITH.ALL;
use IEEE.STD_LOGIC_UNSIGNED.ALL;
entity cnt is
  port ( clk: in std_logic;
       CE : in std_logic;
      reset : in std_logic;
       Counter: out std_logic_vector( 3 downto 0)) ;
end cnt;
architecture Behavioral of cnt is
signal temp_b: std_logic_vector( 3 downto 0);
begin
  process (clk, reset, CE)
  begin
    if reset = '1' then
        temp_b <= "0000";
elsif clk = '1' and clk'event then
      if CE = '1' then
        temp_b <= temp_b + 1;
      end if;
    end if;
  end process;
  counter <= temp_b;
end behavioral;
```

In case of the second counter description, the *temp_b* signal was used to increment the counter output (*counter)*. Notice that the *temp_b* signal was declared in the architecture section as opposed to the variable *temp_a* from the first counter example (*temp_a* declared in the process declaration section). Consequently the signal *temp_b* was visible outside the process statement. Thus, the concurrent assignment *counter <=temp_b*, assures that the counter output is the signal *temp_b* incremented each *clk* cycle within the sequential process. The process is triggered by any change in the signals from the sensitivity list. Thus, a *reset* value *'1'* sets the output to *"0000"*, whereas a CE value *'1'* enables the counting each *clk* cycle, as it can be seen on the simulation results from Fig. 1.6.

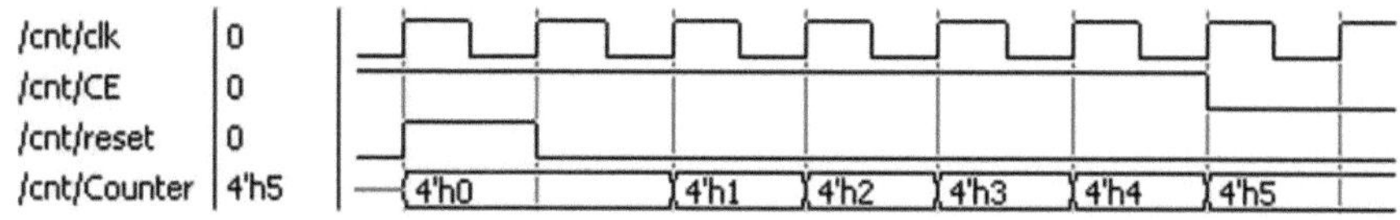

Fig. 1.6 Counter simulation results

If statement

The *if* statement is used for sequential logic description, and, therefore it can be used only inside a process. The syntax of *if* is presented next:

```
if conditions then assignments;
elsif conditions then assignments;
...
else assignments;
end if;
```

The VHDL code for a shift register implementation is provided next for the exemplification of the *if* statement. Moreover, the *generic* statement is used within the shift register in order to define the shift register length. The generic is generally used for the parameterization of the VHDL code. Thus, let us consider we want to build a register which may be used in different logic blocks, and its length may vary depending on the logic block the register is used in. In this case, using generic, the length of the register is once defined as *n* in the entity section (generic (n: integer : = 8);). This *n* value is considered for all the VHDL code description which follows further on. In this way, the VHDL code is easily reusable if we want to describe a shift register of any length, given by *n*. A single change of *n* in the entity section is visible in all the VHDL code. This leads to a parameterized VHDL code for any shift register of length *n* (see the next VHDL code section for exemplification of *generic*).

We describe further on a shift register of variable length *n*. The inputs are *d*, *clk*, *rst* and the output is *q*. The behavior of the circuit is described using the *if* statement; the circuit reset is performed if *'1'* is applied on the *rst* input. Moreover, on rising clock edge (code line: elsif (clk'event and clk = '1')), the input *d* is feed to the *n*'th register cell, whereas the register content is shifted to the right (code line: (temp <= d & temp (n downto 1);). The circuit output *q* is assigned with the first register cell (temp(0)) outside of the register's corresponding process.

Example 1.8—VHDL code for *n* variable length shift register

```
entity shiftreg is
generic (n: integer : = 8);
port (d, clk, rst: in std_logic;
  q: out std_logic);
end shiftreg;
```

```
architecture behavior of shiftreg is
signal temp: std_logic_vector (n downto 0);
begin
  process (clk, rst)
  begin
    if (rst = '1') then
      temp <= (others => '0');
    elsif (clk'event and clk = '1') then
      temp <= d & temp (n downto 1);
    end if;
  end process;
  q <= temp(0);
end behavior;
——————————— end of shift register example ———————————
```

The shift register and its behavior are described in the Figs. 1.7 and 1.8, respectively. As expected, in Fig. 1.8 it can be seen that, the '*1*' logic input of the circuit is available at the output with an $n = 8$ clock cycles delay.

Wait statement

The wait statement is also part of the sequential VHDL code. Placed inside processes, the wait statement comes with a specific request concerning the process that is used in. The process cannot have a sensitivity list. The syntax for the wait statement has three forms, mentioned further on:

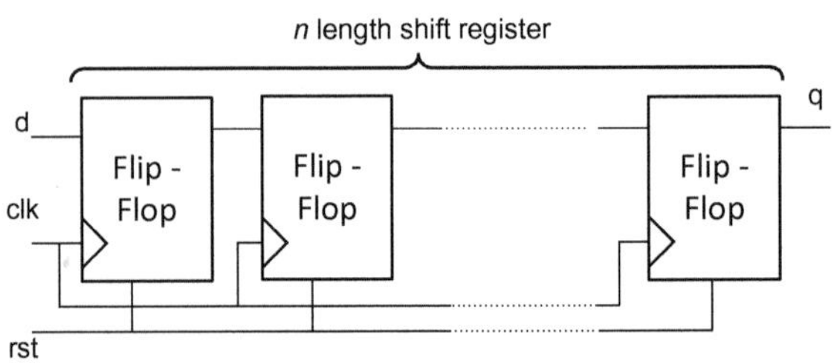

Fig. 1.7 Counter simulation results

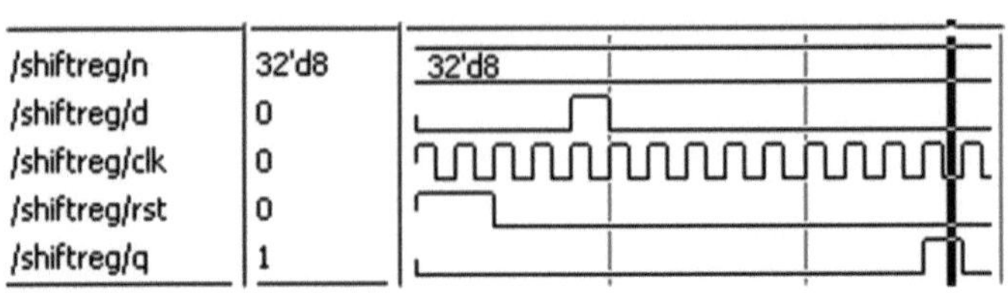

Fig. 1.8 Shift register simulation results

(1) `wait until signal_condition;`

(2) `wait on signal1 [, signal2, ... ];`

(3) `wait for time;`

The first wait statement syntax accepts only one signal condition, and it s mandatory to place the statement at the beginning of the process, since as mentioned before, the process has no sensitivity list. This process is executed when the wait condition is met.

The second wait statement involves multiple signals, any change on the *signal1, signal2,* etc. signals list causes the process execution.

The last wait statement, *wait for time*, is used only for simulation purposes. Thus it is not synthesizable; its use is detailed in the sub-Sect. 1.5 *Writing test-benches.*

1.5 Structural Description with VHDL

Digital systems are commonly described by multiple sub-components and their interconnection links. Like in any other programming languages where a program is composed of multiple sub-programs, a VHDL description of the architecture of a given entity may be hierarchically described by different components (entities) which interact with each other through interconnection links. This type of description is known as a *structural description*. Further on we will discuss how the declaration and the instantiation of a given component is done within the architecture body of an entity that uses structural description.

We've seen so far that, an entity specifies the input output ports of a digital circuit, whereas the circuit behavior is described in the architecture body. In the example what follows, structural VHDL code is used to describe the behavior of the entity named *structural_description*. Two components, *comp1* and *comp2* are declared within the declarative section of the entity's architecture. The syntax for component instantiation is:

```
component comp is port ( port name : signal mode signal_type;
...);
```

Within the architecture body section, the components *comp1* and *comp2* are instantiated. Through instantiation, the input and output signals of the entity *structural_description* are assigned to the components ports. Internal interconnection links between components are drawn using the declared signals such as *w1* signal declared as: "signal w1: std_logic;". The syntax for component instantiation is given next:

```
inst1: comp port map (signals list);
```

As it can be noticed, a labeled is used for component instantiation (*inst1*) followed bt the component name (*comp*), *port map* identifier and the *signals list* in between brackets. The signals found in the *signal list*, associate signals to all of the component's input and output ports.

For the exemplification of components declaration and instantiation, let us consider the logic circuit from Fig. 1.9. Here, the logic circuit named *structural_description,* having 3 inputs ports A, B, C and an output X, is illustrated. The circuit behavior is described by two components *comp1* and *comp2*, and their interconnection using the wire *w1*. In order to describe this circuit using VHDL code, a structural description is used (see the next VHDL code section *VHDL code for structural description*). An entity entitled *structural_description* is used to define the input and output ports of our circuit (A, B, C, D and X). Two components *comp1* and *comp2* are declared in the architecture section together with a signal *w1* used for interconnection link between the two components. After the declaration section the architecture description begins. Here, the two components are instantiated. The instantiation assigns the input and outputs for each of the components, using the ports of the *structural_description* entity and the interconnection wires defined as signals (e.g. *w1* signal).

Example 1.9— VHDL code for structural description

```
entity structural_description is
Port (A, B, C: in STD_LOGIC;
       X: out STD_LOGIC);
end structural_descripton;
architecture Behavioral of structural_description is
component comp1 is port (C1, C2: in std_logic;
   Out1: out std_logic);
end component comp1;
component comp2 is port(C3, C4 : in std_logic;
   Out2: out std_logic);
end component comp2;
signal w1: std_logic;
begin
  inst1: comp1 port map (A, B, w1);
  inst2: comp2 port map (w1, C, X);
end Behavioral;
```

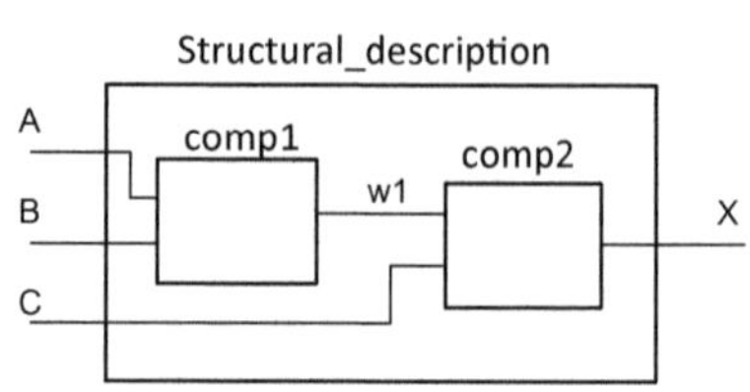

Fig. 1.9 Example of a structural description

The previous example can be used as the starting point to describe more complex digital circuits using structural description. Consequently, a digital logic circuit is described next using VHDL structural description, in order to accomplish an image processing task, image profile computation. In the next section the computation of image profiles is described, followed by the description of the digital logic which performs the computation.

A digital image is commonly represented as a two dimensional array of intensities (pixels) denoted by $I = (p_{x,y})$, where p_{xy} is the pixel intensity values for the pixel found at the *(x,y)* location. In case of a color image denoted by I_c, a given pixels *(x,y)* is described by a set of 3 intensity values $(r_{x,y}, g_{x,y}, b_{x,y})$, corresponding to the colorimetric information red (R), green (G) and blue (B), respectively. The luminance information in case of a color image I_c is obtained as a weighted sum of the R, G and B intensities for all *(x,y)* image pixels as in Eq. (1.2).

$$lum_{x,y} = 0.299 \cdot \mathrm{r_{x,y}} + 0.587 \cdot \mathrm{g_{x,y}} + 0.114 \cdot \mathrm{b_{x,y}} \tag{1.2}$$

Summing up pixel intensities representing luminance information along *x* and *y* image direction lead to the vertical and horizontal luminance function profiles of an image, as expressed by Eqs. (1.3) and (1.4), respectively.

$$V(y) = \sum_x lum_{x,y} \tag{1.3}$$

$$H(x) = \sum_y lum_{x,y} \tag{1.4}$$

The overall view of the corresponding digital logic for image profile computation is illustrated in Fig. 1.10. The logic blocks used are the *RGB to Luminance*, the *Address Computation Unit*, and the *Accumulators* for each of the profiles, vertical and horizontal one.

The image colorimetric information is delivered pixel-wise as input to the *RGB to Luminance* logic block, which delivers the luminance information for each *(x,y)* image pixel. The *clk* and the *start* inputs are also present for the *RGB to Luminance*

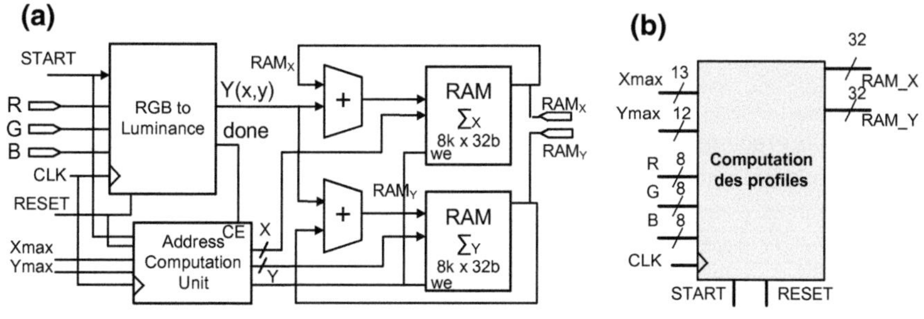

Fig. 1.10 Block diagram for image profile computation

logic block. The *start* input marks the presence at the input of each RGB triplets for which the luminance is computed. Note that, according to Eq. (1.2), for one Y value computation, 3 multiplications and 2 additions are needed. A number of two clock cycles are needed, thus the output *Y* of the circuit is available with 2 *clk* cycles latency. The data ready on the output port is signalized by the *done* output, which represents the *start* input propagated through the *done_pipe* register. The size of the *done_pipe* register is given by the constant *pipe_level* set to 1. Further on we will focus on the luminance computation. The proposed logic block takes advantage of the fact that both luminance and colorimetric information are represented as integer values. Thus, *R, G and B* values are scaled, by multiplication with a factor of 2^8. The scaling is performed in order to have integer factors in the Eq. (1.2). The final results is then divided by 2^8, by means of bits shifting (code line *Y = temp (15 downto 8)*), in order to have the luminance output Y in the range of 1 to 255. The VHDL code used for the description of previous functionality is provided in the next VHDL code section.

Example 1.10— VHDL code for RGB to luminance conversion

```
entity RGBtoY is
   generic (c : integer : = 8);
port (clk: in std_logic;
         start: in std_logic;
       R: in std_logic_vector (7 downto 0);
       G: in std_logic_vector (7 downto 0);
       B: in std_logic_vector (7 downto 0);
       Y: out std_logic_vector (7 downto 0);
         done : out std_logic);
end RGBtoY;
architecture behavioral of rgbtoy is
constant r_coef: unsigned (7 downto 0) : = to_unsigned( integer( real
(0.299)
     * real(2**(c)-1) ), c);- 0.299 * 255
constant g_coef: unsigned (7 downto 0) : = to_unsigned( integer( real
(0.587)
     * real(2**(c)-1) ), c);- 0.587 * 255
constant b_coef: unsigned (7 downto 0) : = to_unsigned( integer( real
(0.114)
      * real(2**(c)-1) ), c);- 0.114 * 255
signal r_scaled, g_scaled, b_scaled: unsigned (8 + c-1 downto 0);
constant pipe_level : integer: = 1;
signal done_pipe: std_logic_vector
(pipe_level-1 downto 0): = (others => '0');
signal temp : std_logic_vector (15 downto 0);
begin
process (clk)
```

```
begin
   if clk'event and clk = '1' then
        r_scaled <= unsigned(R) * r_coef;
        g_scaled <= unsigned(G) * g_coef;
        b_scaled <= unsigned(B) * b_coef;
    temp <= std_logic_vector(r_scaled + g_scaled + b_scaled);
        done_pipe(0) <= start;
        done <= done_pipe(0);
    end if;
end process;
y <= temp(15 downto 8);
end behavioral;
—————————————— End of RGB to luminance conversion ————
```

As part of the profile computation architecture schematic (see Fig. 1.9), the two *accumulators* used to compute the image vertical and horizontal profile are discussed. Pixel-wise luminance information Y is delivered to both accumulators. Using an adder for each memory, a specific memory value x is read from the memory and the current pixel luminance is added to the x value. This approach leads to the completion of the content for the two memories, which in the end represent the image profiles. The next VHDL code section infers a Block RAM memory for horizontal image profile storage. The inferred Block RAM is called *RAM_X*. Further details on memory inference and instantiation are provided in Chap. 6.

Example 1.11— RAM_X memory inference for horizontal profile computation and storage

```
entity RAM_X is
 Port (CLK : in STD_LOGIC;
          we: in std_logic;
          addr: in std_logic_vector(12 downto 0);
          data: in std_logic_vector (31 downto 0);
          data_out: out std_logic_vector (31 downto 0));
end RAM_X;
architecture Behavioral of RAM_X is
type RAM is array (2**13-1 downto 0) of std_logic_vector (32-1 downto 0);
signal RAM_X: RAM: = (others => (others => '0'));
begin
process (CLK)
begin
  data_out <= RAM_X(conv_integer(addr));
  if (CLK'event and CLK = '1') then
    if (we = '1') then
      RAM_X(conv_integer(addr)) <= data;
```

```
    end if;
  end if;
end process;
end Behavioral;
------------------ End of block RAM memory inference ------------------
```

The functionality of the Block RAM is summarized as follows: the input *data* is written at the *addr* memory address in case the write enable input *we* is *1* logic. A concurrent statement (i.e. data_out <= RAM_X(conv_integer(addr));) delivers the memory content from the *addr* address to the output *data_out*, performing a memory read operation. In order to build an accumulator using this type of memory for both horizontal and vertical image profiles, two adders are inferred one for each memory. The adder inputs at a given time are, the current pixel luminance information and the memory values read from a *specific memory address.* The adder output represents a sum to be written on the same memory location from which the first read was done.

The question that arises is which are the *specific memory addresses* to be given as inputs to the block RAM memories, in order to build the image profile whereas all image pixels are delivered sequentially at the input. The logic block *address computation unit (ACU)* answers this question, by providing the *x* and *y* addresses to the block ram memories for the horizontal and vertical image profile, respectively. The VHDL code for the *ACU* logic block is VHDL code section.

Example 1.12— Address computation unit

```
entity x_y is
Port  (Xmax :in std_logic_vector (12 downto 0);
      Ymax : in std_logic_vector (11 downto 0);
CLK : in  STD_LOGIC;
Reset: in STD_LOGIC;
      Start : in std_logic;
      X : out STD_LOGIC_VECTOR (12 downto 0);
Y : out STD_LOGIC_VECTOR (11 downto 0);
      done : out std_logic);
end x_y;
architecture Behavioral of x_y is
signal x_out : UNSIGNED(12 downto 0) : = to_unsigned (0,13);
signal y_out : UNSIGNED(11 downto 0) : = to_unsigned (0,12);
signal reset_v : std_logic: = '0';
signal first_start : std_logic: = '0';
signal done_pipe: std_logic_vector (1 downto 0): = (others => '0');
begin
process (CLK, Reset)
begin
  if reset = '1' then
```

```
x_out <= (others => '0');
y_out <= (others => '0');
reset_v <= '1';
     elsif CLK = '1' and CLK'event then
           if (start = '1' and first_start = '1') or reset_v = '1' then
                 if x_out < UNSIGNED(Xmax) then
                            x_out <= x_out + 1;
                      elsif y_out < UNSIGNED (Ymax) then
                            x_out <= (others => '0');
                            y_out <= y_out + 1;
                      else
                            x_out <= (others => '0');
                            y_out <= (others => '0');
                            reset_v <='0';
                      end if;
                 end if;
done_pipe(0) <= START;
           done_pipe(1) <= done_pipe(0);
           if done_pipe(1) = '1' then
           first_start <= '1';
           end if;
     end if;
end process;
done <= done_pipe(1);
X <=std_logic_vector(x_out);
Y <= std_logic_vector (y_out);
end Behavioral;
----------------- End of address computation unit -----------------
```

The input /output ports of the ACU included in the entity declaration section are: (i) *Xmax* and *Ymax* inputs, corresponding to the image size, and to the horizontal and vertical profile length, respectively, (ii) the *X* and *Y* outputs corresponding to the memory addresses to be read and written for building up the image profiles, (iii) *start* input which marks the computation start for the current pixel, (iv) the output *done* which signalizes that the addresses are available for the block RAM memories, (v) the *reset* which initializes the output addresses with *0* values. The functionality of the ACU unit is detailed as follows: the 1 logic reset values initializes the internal x_out and y_out registers with 0. The x_out and y_out registers content is concurrently delivered at the ACU output (e.g. X <=std_logic_vector (x_out);). The type cast operator *std_logic_vector* is needed, since the *unsigned* type is used for registers content, since addition operation (incrementing) is used on these registers. Each time an RGB input pixel is available at the overall *image profile computation* architecture, the *start* signalizes this event, and the *x_out* and *y_out* registers values are incremented. Moreover, the *start* input is passed through an two cells internal shift register namely *done_pipe*. This operation is performed in order to allow the RGB conversion block to perform the computation. Thus, with a 2 clock cycles delay, the *done* output marks the availability of both the addresses and luminance information at the memory inputs for profile completion.

Once the functionality for each logic block is described, VHDL structural description is used for interconnecting the components of the image profile computation architecture. The VHDL code section corresponding to the structural description of the proposed architecture is detailed in the Appendix B. Each component (i.e. *RGB to Luminance*, *Address Computation Unit* and the *Accumulators*) is instantiated and, using signal declarations, the components are interconnected.

In the next section the VHDL language constructs and semantics for building test-benches aiming to test the functionality of digital logic blocks are described. Prior to VHDL language for test-benches description, a test-bench is build for the proposed architecture for image profile computation.

1.6 VHDL Code for Simulation Test-Benches

The current section presents how the functionality of digital logic blocks (i.e. units) is verified by means of simulation. For simulating the functionality of a digital logic block, test-benches are developed using VHDL language. Simulation test-benches generally need two inputs: the description of the design unit called unit under test (UUT) and stimulus description to drive the design. The top level design for a simulation test-bench is illustrated in Fig. 1.10. Note that the test-bench design has no connection to "the outside world". It contains only internal signals to connect the two instantiated components: (1) the unit under test (UUT) and (2) the stimulus driver. Thus, the entity declaration for a test-bench used to simulate the functionality of the UUT is:

```
——— UUT —————
entity test-bench is
-empty
end test-bench;
——— end of UUT ———
```

When writing test-benches, it is essential to understand the difference between VHDL code for the two test-bench components, the unit under test and the stimuli driver. Thus, synthesizable VHDL code constructs are used to describe the unit under test. In this way, the synthesis tool can generate the digital logic with the desired functionality for the unit under test. Further on, stimuli with correct timing are applied to the inputs of the UUT and the outputs are checked using internal test-bench signals. These two operations (i.e. applying stimuli and checking outputs) are performed using the stimuli driver. The VHDL code for stimuli driver description is non-synthesizable (Fig. 1.11).

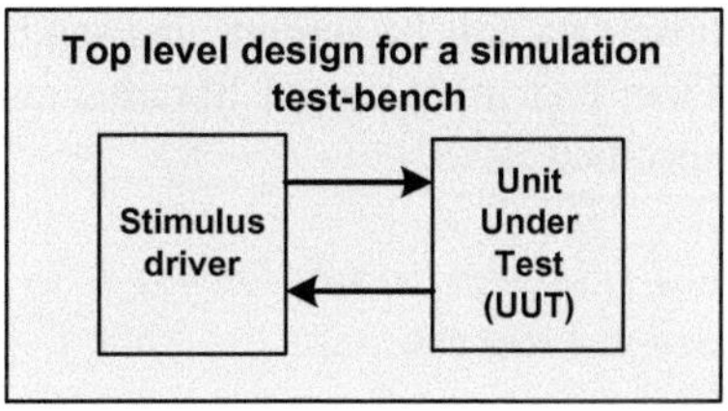

Fig. 1.11 Block diagram for image profile computation

We've seen so far that synthesizable and non-synthesizable VHDL code constructs correspond to digital design unit description and stimuli driver description, respectively. Consequently, it comes natural for the data types to be also classified as synthesizable and non-synthesizable. An eloquent example for non-synthesizable data type is the type *file*. This allows input signal values to be stored in an external text file. This file can be read and its content (e.g. integer values) can be delivered as stimuli to the input signals of the UUT using the stimuli driver unit. To have a view on all predefined VHDL data types and their classification as synthesizable or non-synthesizable, the Table 1.2 is presented.

A special attention is given next to the *time* and the *file* data types. Considering time, in case of synthesizable VHDL code, only time-less assignments are used. On the other hand, for test-bench description, the assignments can be time driven; this means the moment the assignment takes place can be specified using *time*. An example is given next:

```
y <= '1' after 1 ns;
```

Relative to the current simulation time, the *y* signal assignment is performed after *1 ns*. The units that can be used for a variable of *time* data type are: *fs* (femtosecond), *ps* = 1000 *fs* (picosecond), *ns* = 1000 *ps* (nanosecond), *us* = 1000 *ns* (microsecond), *ms* = 1000 *us* (millisecond), *sec* = 1000 *ms* (second), *min* (minute), *hr* (hour).

As referred to the *file* data type, it is used to deliver input test vectors to the unit under test from a given file. Moreover, file data types can store simulation results for further analysis. Samples of VHDL code are provided next in order to illustrate

Table 1.2 Predefined data types included in the standard library

(use std.standard.all)	
Synthesizable data types (used for digital logic description)	*Non-synthesizable data types (used for stimuli driver description)*
Bit	*character*
Boolean	*String*
Int3eger, natural positive	*Real*
bit_vector	*time*

how data can be written and read from text files by the simulation test-benches. First, text input output libraries must be included in the test-bench description as specified next.

```
use IEEE.STD_LOGIC_TEXTIO.ALL;
use STD.TEXTIO.ALL;
```

All files to be used in the test-bench must be declared using the *file* statement. The declarations are included in the test-bench architecture or in the main test-bench process. The supported modes for opening the files are *read_mode*, *write_mode* or *append mode*. These file opening modes are also specified within the file declaration section as it can be seen in the next code example.

```
file my_file1: text open read_mode is "inputdata.txt";
file my_file2: text open write_mode is "results.txt";
```

Once the files declared, *readline* function can be used to hold the data elements from file lines. Thus, a variable *in_line* is declared as one of the file lines and it is followed by the function which performs the file line read operation into the declared variable:

```
variable In_line: line;
readline(my_file1, In_line);
```

One file line includes one or more fields. Individual fields in each line of file are further on accessed by a *read* procedure, defined in the Text IO package. Each read procedure call includes a set of parameters which include the file line, a destination variable to which the value for each line field are assigned, and a boolean variable which checks if the line field value corresponds to the type of the variable used to store the individual line fields values. The VHDL types supported for the destination variable of the read procedure are bit, bit_vector, boolean, character, string, integer, and time. An example of a read operation is provided next, where an integer variable *destination* is used to read the fields of the *In_line* variable, representing one line of the *my_file1* file.

```
variable destination: integer;
variable check: Boolean;
read (In_line, destination, check);
```

Readline and *read* statements can be executed until an end-of-file or an end-of-line condition, respectively. Thus, a loop within the test-bench main process can be used to verify these conditions and to perform sequential *read* or *readline* operations until the conditions are fulfilled.

```
while not endfile(my_file1) loop
readline (my_file1, In_line);
...
end loop;
```

The VHDL simulation mechanism

The simulation mechanism is based on multiple processes which drive stimuli to the inputs of the unit under test and also check the outputs of the same unit. There are three steps involved in the VHDL simulation mechanism: (1) *elaboration*, (2) *initialization* and (3) *execution*. In the first step, all the design units, stimuli driver and the units under test are compiled and loaded for simulation. In the second step, all *drivers* (i.e. variable and signals) are given their initial values or default values in case no initial value is present. By drivers we understand signals and variables used to deliver input data and to read the output data to and from the unit under test. Once the initial values are set, the *execution step* starts.

The execution step is a cyclic process driven by events /changes that occur on the test-bench signals/drivers. Once the changes occur in the signals, the test-bench processes are entering the execution phase. One simulation cycle is known as *delta cycle*. The execution step is divided into two phases: *(i) process execution phase* and *(ii) signal update phase*.

In the *process execution phase* all active processes are executed. Variables immediately assigned and signals are scheduled for assignment in the next signal assignment phase.

In the *signal update phase*, signals are updated. This update operation can activate other processes or even the same process, due to the sensitivity to the changed signals. For a better understanding, the *execution step* together with its two phases is discussed in case of two processes described in the next VHDL code section.

```
——— first process ———
Gate: process(A,B)
begin
S <= A and B;
End process;
——— second process ———
Inverter: process (S)
Begin
Z <= not S;
End;
```

Considering the two processes, we assume that the current simulation time is 10 ns. In case a change in the *A* signal occurs, the first process enters the *process execution phase*, and the signal *S* is scheduled for assignment in the next *signal*

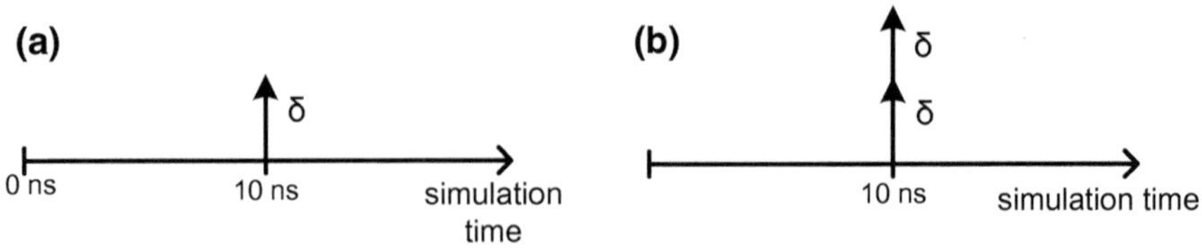

Fig. 1.12 The simulation time and the delta cycle of the simulation process

update phase. In terms of delta cycles, the update of the signal *S* is scheduled to change at *10 ns + delta*.

After the execution phase the *signal update phase* comes and the *S* signal is updated at *10 ns + delta* (Fig. 1.12a). This activates the second process which enters the execution phase and the *Z* signal is scheduled to change at the next signal update phase at *10 ns + 2·delta*. At this time no processes are activated since no change are on the sensitivity lists, so the simulator time increases.

Example of VHDL test-bench

Considering the image profile computation unit described in Fig. 1.10, the VHDL code for the test-bench used for its simulation is described next. As previously discussed, no inputs or outputs are present in the test-bench entity. On the other hand, the architectior declaration sections includes the component which performs the image profile computation, called *Prof_comp*. The declaration section contains also the signals which drive the inputs and outputs values to the UUT. The list of signals include the size of the profiles (*Xmax, Ymax*), *clock*, the *R, G* and *B* pixel intensity values for the 3 image channels, the *start* which marks new pixel intensity values present at the input, the *X* and *Y* outputs which monitor the read and write addresses generated for the Block RAM memories which store the profiles, the *DataRAM_X* and *DataRAM_Y* which monitor the read and written data into the Block RAMs, and last the *done* signal which marks pixel intensity data has been added into both the Block RAMs corresponding to the vertical and horizontal image profiles. Considering test-bench architecture, it contains the instantiation of the unit under test (UUT) and two processes, one for the clock generation and one process for stimuli driver to the UUT.

Example 1.13—Test-bench for the image profile computation unit

```
LIBRARY ieee;
USE ieee.std_logic_1164.ALL;
USE ieee.std_logic_unsigned.all;
USE ieee.numeric_std.ALL;
ENTITY test_all IS
END test_all;
ARCHITECTURE behavior OF test_all IS
- Component Declaration for the Unit Under Test (UUT)
```

```
COMPONENT Prof_comp
   PORT(
      Xmax : IN std_logic_vector(12 downto 0);
      Ymax : IN std_logic_vector(11 downto 0);
      CLK : IN std_logic;
      Start : IN std_logic;
      Reset : IN std_logic;
       R : IN std_logic_vector(7 downto 0);
      G : IN std_logic_vector(7 downto 0);
      B : IN std_logic_vector(7 downto 0);
   X : OUT std_logic_vector(12 downto 0);
      Y : OUT std_logic_vector(11 downto 0);
      DataRAM_X : out std_logic_vector (31 downto 0);
      DataRAM_Y : out std_logic_vector (31 downto 0);
      DBG_we : out std_logic;
      DBG_sum_x : out std_logic_vector (31 downto 0);
      done : OUT std_logic
      );
   END COMPONENT;
  -Inputs
signal Xmax : std_logic_vector(12 downto 0) : = (others => '0');
signal Ymax : std_logic_vector(11 downto 0) : = (others => '0');
signal CLK : std_logic : = '0';
signal Start : std_logic : = '0';
signal Reset : std_logic : = '0';
signal mux_mode : std_logic;
signal addr_in : std_logic_vector(31 downto 0);
signal R : std_logic_vector(7 downto 0) : = (others => '0');
signal G : std_logic_vector(7 downto 0) : = (others => '0');
signal B : std_logic_vector(7 downto 0) : = (others => '0');
-Outputs
signal X : std_logic_vector(12 downto 0);
signal Y : std_logic_vector(11 downto 0);
signal DBG_we : std_logic;
signal DBG_sum_x : std_logic_vector(31 downto 0);
signal reset_out : std_logic;
signal DataRAM_X : std_logic_vector(31 downto 0);
signal DataRAM_Y : std_logic_vector(31 downto 0);
constant CLK_period: time : = 10 ns;
BEGIN
- Instantiate the Unit Under Test (UUT)
  uut: Prof_comp PORT MAP (
       Xmax => Xmax,
       Ymax => Ymax,
       CLK => CLK,
```

```
        Start  => Start,
        Reset  => Reset,
     R  => R,
        G  => G,
        B  => B,
        X  => X,
        Y  => Y,
         DBG_we  => DBG_we,
        DBG_sum_x  => DBG_sum_x,
        DataRAM_X  => DataRAM_X,
        DataRAM_Y  => DataRAM_Y,
        done  => reset_out
      );
- clock generation process
  CLK_process : process
  begin
   CLK  <= '0';
   wait for CLK_period/2;
   CLK  <= '1';
   wait for CLK_period/2;
  end process;
- Process to drive stimuli to the UUT
  stim_proc: process
  begin
        xmax  <= '0'&X"004";
        ymax  <= X"005";
        wait for 10 ns;
        wait until CLK'EVENT and CLK = '1';
        reset  <= '1';
        wait until CLK'EVENT and CLK = '1';
        reset  <= '0';
        wait until CLK'EVENT and CLK = '1';
        wait for 80 ns;
        R <= X"25" after 2 ns;
        G <= X"25" after 2 ns;
     B <= X"25" after 2 ns;
        start  <= '1';
          wait until clk'event and clk  =  '1';
          start  <= '0' after 2 ns;
        wait for 5 ns;
        wait until CLK'EVENT and CLK = '1';
        wait for 80 ns;
        R <= X"AA" after 2 ns;
      · G <= X"AA" after 2 ns;
        B <= X"AA" after 2 ns;
```

```
        start <= '1';
        wait until clk'event and clk = '1';
        start <= '0' after 2 ns;
        wait for 5 ns;
       wait until CLK'EVENT and CLK = '1';
        wait for 80 ns;
        R <= X"05" after 2 ns;
        G <= X"05" after 2 ns;
        B <= X"05" after 2 ns;
        start <= '1';
        wait until clk'event and clk = '1';
        start <= '0' after 2 ns;
        wait for 5 ns;
        wait until CLK'EVENT and CLK = '1';
        wait for 80 ns;
   wait;
   end process;
END;
---------------------- End of test bench ----------------------
```

The stimuli driver process assigns input values and reads outputs to and from the unit under test ports using internal signals. As it can be seen in the simulation from Fig. 1.13, after a reset is applied at the UUT input after *10 ns*, first R, G, B input is feed to the profile computation unit after another *80 ns*, at the t_0 simulation time. The luminance information for the current pixel is computed according to Eq. (1.2) and is given with a 2 clock cycles delay (t_1 simulation time). After another 2 clock cycles

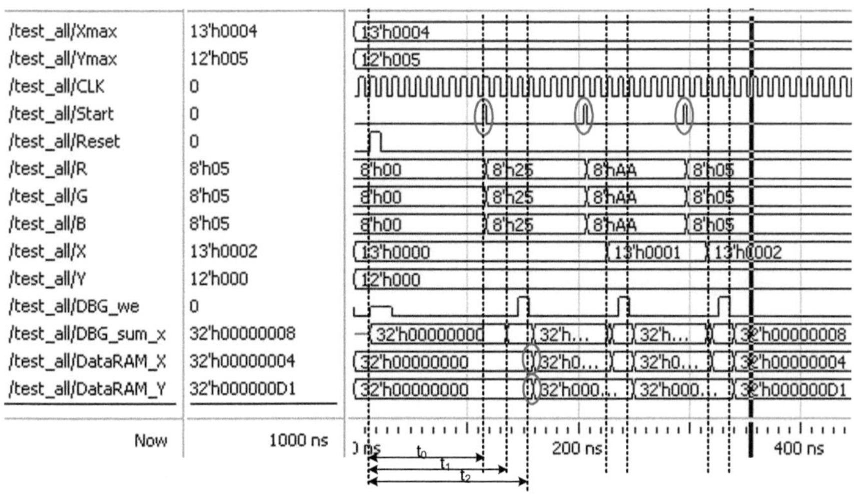

Fig. 1.13 Simulation of write operations in the Block RAM memories for the completion of the image vertical and horizontal profiles

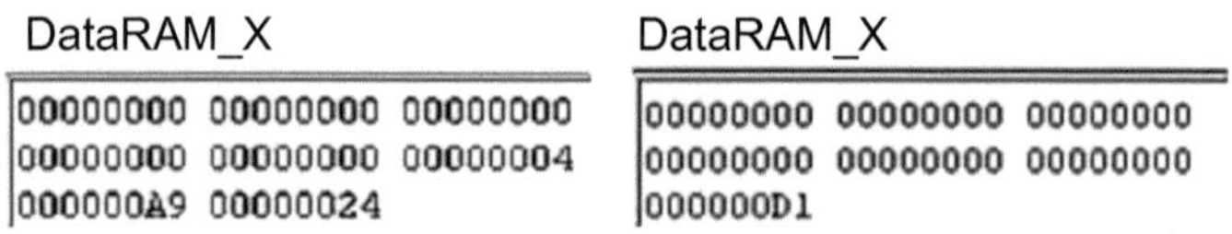

Fig. 1.14 Block RAM memory contents corresponding to the image vertical and horizontal profile, considering the simulation from Fig. 1.12

delay (t_2 simulation time), the current memory value is read and the luminance intensity value is added for the current pixel at the memory location *X* and *Y* from the two Block RAM memories. For the first pixel intensity value, the addresses *X* and *Y* are not incremented (the values are *0* and *0*). The address incrementing for the horizontal profile starts with the second image pixel, and is marked with a rectangle on the simulation from Fig. 1.13. The address for the vertical image profile is incremented only after the first line of pixels is feed to the UUT.

The content of the block RAM memories corresponding to the image profiles is given in the Fig. 1.14, according to the write operations performed in the simulation from Fig. 1.13. In the memory which corresponds to the vertical profile (*DataRAM_X*), each of the three intensity values corresponding to the three R, G, B inputs is written at the addresses from *0* to *2*. In case of the vertical image profile, each of the three current pixel intensity values from the input are sequentially added to the content located at the first memory address. In this way, considering the memory which stores the vertical image profile (*DataRAM_Y*), the sum of the three pixel intensity values corresponding to the three sets of R,G,B inputs is found at the first memory address at the end of the simulations from Fig. 1.13.

1.7 Finite State Machines

A *finite state machine* (FSM) can be defined as an abstract computation procedure described by a sequence of steps. The FSM can be in one of a finite number of states at a given time. Depending on its current state and a sequence of events, the procedure passes from one state to another (i.e. state transition). Each state is associated with a set of operations, and consequently, the finite state machine describes a sequence of operations depending on the events that occur in each of its finite states [3].

As previous definition underlines, the FSM controls how a sequence of operations is performed. From the perspective of digital logic, the operations are performed by an *execution unit*, whereas the timing and the conditions to be fulfilled in order to activate the execution unit are controlled by the control unit (i.e. controller). The execution is composed of registers, adders, decoders, comparators or arithmetic logic units (ALU) and it usually performs arithmetic operations, logic operations, data processing tasks and also interprets control signals and generate status signals for the control unit. The control unit on the other hand, controls data movement by disabling and enabling resources, provides signals to activate tasks

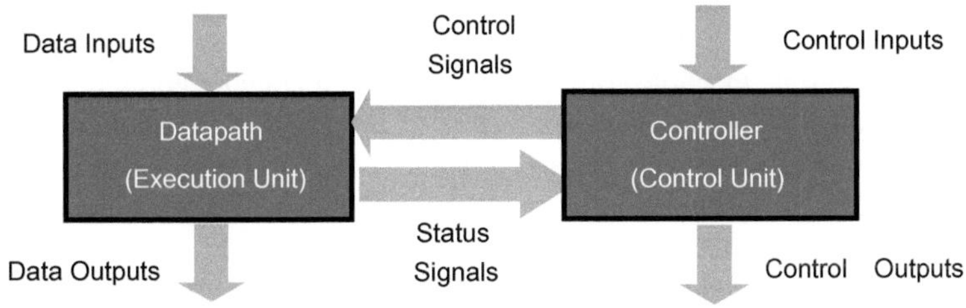

Fig. 1.15 Execution Unit and Control Unit in case of digital systems

for the execution unit, determines the timing (i.e. sequence of operations in time) of the execution unit operations and also interprets the received signals from the execution unit and generate signals for the execution unit. An overview of how the two units interact is given in Fig. 1.15.

Commonly the control unit can be described by a finite state machine. The digital circuit which implements the functionality of a FSM is composed of the following components: (i) a state register, (ii) input signals, (iii) a next state logic function, (iv) the output function and (v) the output signals. The *state register* is used to encode all the symbolic states of the FSM. The *next state function* determines the new state the FSM transits to from the current state, based on the current state and the input signals. The *output function* specifies the output signal values depending only on the current state or on both current state and input values.

Based on how the output is computed, the FSM are classified asof type Moore or type Mealy. In case the output depends only on the present state of the FSM, the FSM is of type Moore, whereas in case the output depends only on the input signals the FSM is of type Mealy. The logic diagrams for both Moore and Mealy FSM are presented in Fig. 1.16.

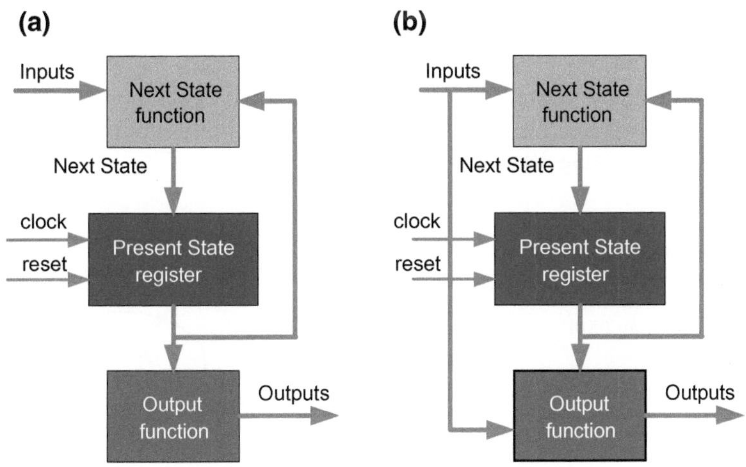

Fig. 1.16 State transitions in case of a simple FSM

State diagrams for FSM representation

Designing finite state machine commonly starts with a graphic representation of the FSM states called *state diagram*. Each state of the FSM is represented as a *node* drawn as a circle together with unidirectional *arrowed arcs* which enter or exit the state node. Each arc has associated a condition. Thus the arcs mark the transitions from one state to another in case the arc condition is met. Moreover, the outputs assignment is represented for each of the FSM states. In case of a Moore FSM, the outputs assignments are placed inside the circle of the state, whereas in case of a Mealy FSM, the assignments are placed under the condition of each arc. This representation, detailed in Fig. 1.17, incorporates all information which characterizes an FSM state: state name, input, output, next state function (i.e. conditional transitions) and also the output function. The FSM passes from *State A* to *State B* in case the condition is fulfilled. Once in *State B*, the output of type Moore is assigned (*Moore_out* <= *value;*) and the FSM passes to one of the next states, depending on conditions 1.1 and 1.2. In case *Condition* 1.2 is fulfilled, the Mealy outputs are assigned based on both the condition 1.2 and the input signals.

For a better understanding an example of a FSM state machine for a memory controller is presented next. Thus, the input/output ports together with the functionality of a memory controller are described by the logic block from Fig. 1.18a and the state diagram from Fig. 1.18b, respectively.

The memory controller logic block is characterized by the *rw*, *ready*, *reset* and *clk* inputs and also by the outputs *oe* and *we*. The *reset* input signal having the logic value '1' brings the FSM in the state IDLE. In this state, in case the memory is ready for a read or write operation (*ready* = *'1'*), the FSM passes to the *Decision* state. In this state, based on the *rw* input signals, the FSM passes to one of the *Read* or *Write* states, where the *oe* and *we* outputs are assigned. Thus, in the state *Read*, the *oe* signals is assigned with '0' logic value, whereas the *we* signal is assigned to '1' logic value. On the other hand, *oe* is assigned with '1' logic value and *we* with '0' logic value in case of the state *Write*. Once in the *Read* or *Write* state, if the current memory operation is finalized, the *ready* input marks this with '1' logic value which brings the FSM in the *IDLE* state. This means the FSM is ready for another read or write operation. Having both the input/output ports and the FSM functionality described, the next step in memory controller implementation is to write the corresponding VHDL code.

Fig. 1.17 **a** Memory controller logic block and **b** its functionality described by the corresponding FSM

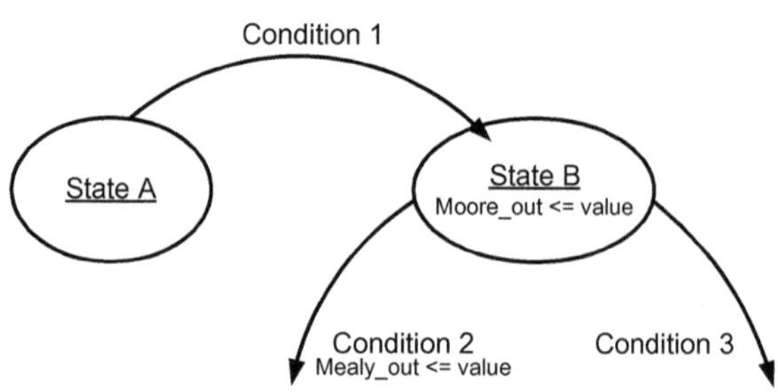

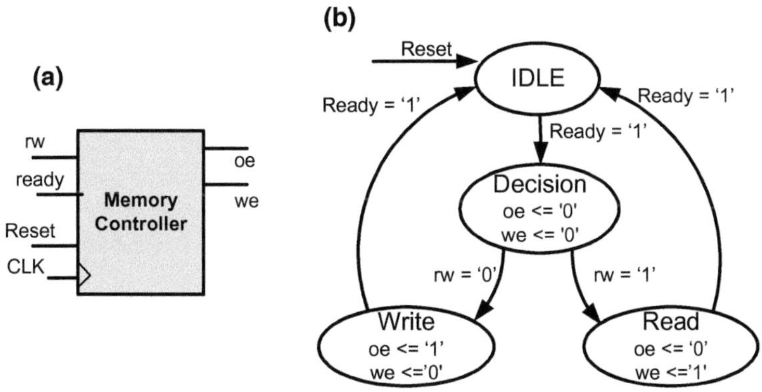

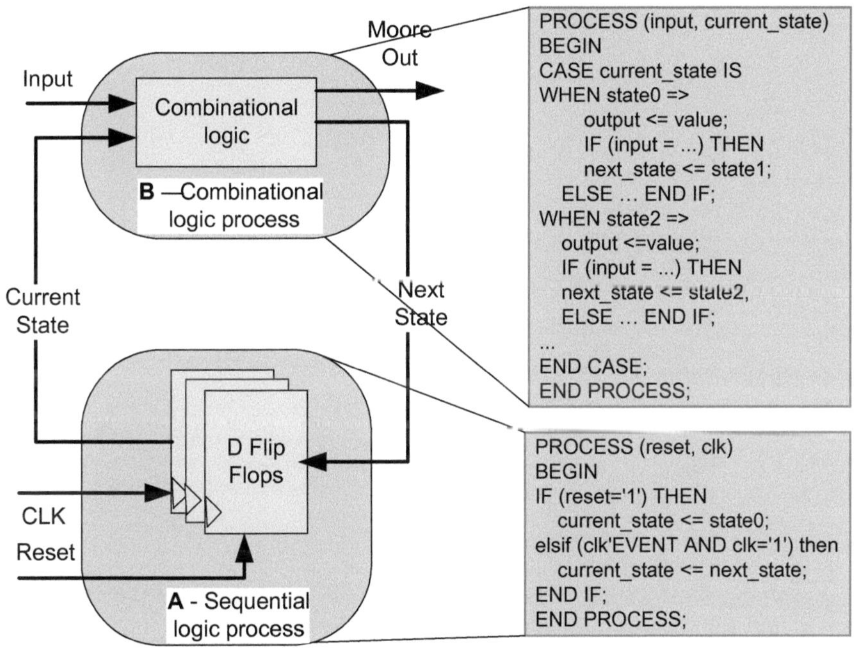

Fig. 1.18 VHDL code composed of 2 processes and the corresponding logic diagram for both the sequential and the combinational logic

VHDL code for FSM implementation

Before getting into the FSM description using VHDL, let us consider the functionality of the FSM detailed in Fig. 1.16. Two types of processes are distinguished: (A) one combinational process for the next state logic function and (B) one combinational process corresponding to the state register and its transitions from one state to another. The transitions are performed depending on the *clk* signal.

The entity declaration together with the signals declaration for the finite state machine is presented in the next code sequence. The two processes, the sequential process A and the combinational process B, which describe the functionality of the FSM, are defined in the architecture body as detailed in Fig. 1.19.

Before discussing the two processes, we focus on the declarative section of the FSM architecture. Here, the type state is defined which includes the stated *state1*, *state2*, *state3*... which correspond to the FSM functionality. Moreover, the signals *current_state* and *next_state* of type *state* are declared. They are further on used in both A and B processes. In the sequential process A, *current_state* is initialized with *state0* upon reset signal and, on risng clock edge the *next_state* is assigned the *current_state* signal value. In the combinational process B, the transitions from all possible states are specified (e.g. *next_state <= state1*). The transitions are performed based on the current state and the inputs values (e.g. *when state = state 0 /if (input = value) then /output < = value; next_state <= state1*;).

```
--------------VHDL code structure for a Moore FSM--------------------
library ieee;
use ieee.std_logic_1164.all;
entity fsm is
port (input: in < data_type > ;
reset, clk: in std_logic;
output: out < data_type >);
end fsm;
architecture Behavioral of fsm is
type state is (state0, state1, state2, state3, ...);
signal current_state, next_state: state;
begin
process A; - these two process are detailed
process B; - in the next figure
end;
```

Regarding the FSM description from Fig. 1.18, the *output* assignment is discussed next. It can be seen the output depends only on the *current_state*, and consequently, we have a Moore FSM. In case the outputs depend also on the circuit input, the type of the FSM is Mealy. In this case, the output changes in case the input changes, which leads to the following description of the combinational process B:

```
----------VHDL code structure for an asynchronous Mealy FSM------------
process (input, current_state)
begin
case current_state is
when state0 =>
  if (input = ...) then
```

```
      output <= value;
      next_state <= state1;
    else ... end if;
when state2 =>
...
end case;
end process;
```

The previous VHDL code describes an asynchronous assignment of the Mealy FSM outputs. In case the signals are required to be synchronous, the output is updated depending on the clock signal edges. In order to be able to do this, the outputs are stored using D flip-flops according to the design from Fig. 1.19. In order to achieve this, the VHDL code to be used is illustrated next.

Example 1.14—VHDL code structure for a synchronous Mealy FSM

```
library ieee;
use ieee.std_logic_1164.all;
entity fsm is
port (input: in < data_type > ;
reset, clk: in std_
: out < data_type >);
end fsm;
architecture behavioral of fsm is
type state is (state0, state1, state2, state3, ...);
signal current_state, next_state: state;
signal temp: < data_type > ;
begin
process (input, current_state)
begin
case current_state is
when state0 =>
    if (input = ...) then
      temp <= value;
      next_state <= state1;
    else ... end if;
when state2 =>
    if (input = ...) then
   temp <=value;
      next_state <= state2;
    else ... end if;
...
end case;
end process;
```

```
process (reset, clk)
begin
if (reset = '1') then
   current_state <= state0;
elsif (clk'event and clk = '1') then
   output <= temp;
   current_state <= next_state;
end if;
end process;
end;
```

The VHDL code for describing both Moore and Mealy FSM are detailed in the previous code sections. For the memory controller implementation illustrated in Fig. 1.18, the corresponding VHDL code is given in Appendix C. The functionality of the FSM is presented in the Fig. 1.19. The simulation shows how a write operation and two successive reads are performed. The state transitions from *IDLE* to *decision* and further on from *decision* to *write* or *read* are also illustrated by visualizing the signal state and its evolution on the time diagram from Fig. 1.20.

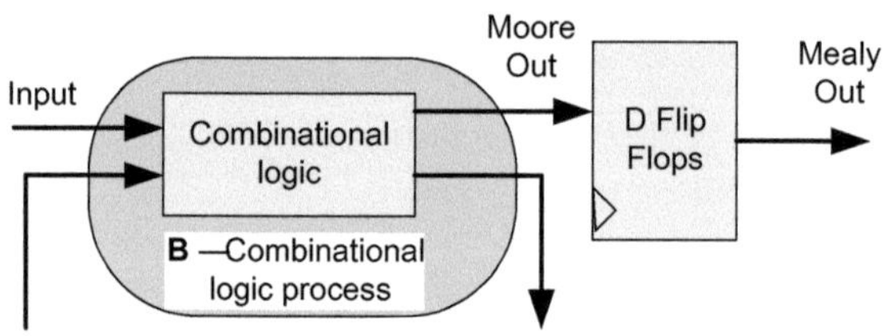

Fig. 1.19 Mealy FSM with synchronous outputs assignment

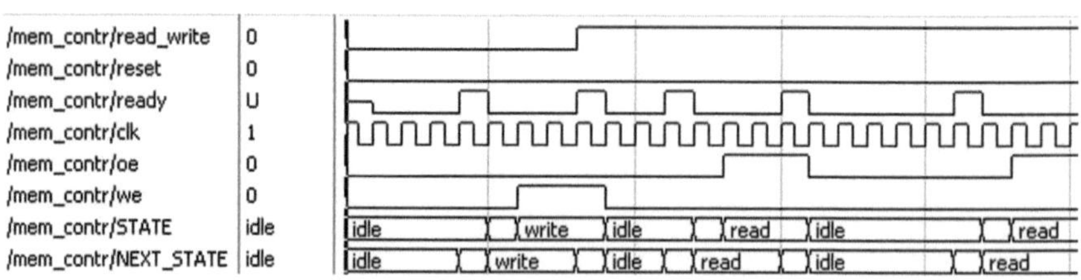

Fig. 1.20 Simulation of the memory controller functionality

1.8 Methodology for Digital Design with VHDL

In the introductory section we have seen how different abstraction levels can be used in describing digital systems. Moreover, a general view on designing digital systems has been also discussed. In this context, processes such as synthesis, physical design, verification and testing have been defined. The current section presents each of these processes from the perspective of a designer which makes use of the VHDL language to build digital systems.

Let us consider a specific algorithm composed of a sequence of steps applied on our input data, which lead to the corresponding outputs, or in other words to the expected results. Our aim is to implement this algorithm using digital logic. The tools available to achieve this task are the VHDL language, a target FPGA together with the specific software packages used to perform the synthesis. Using the VHDL language constructs, the functionality of our algorithm is described using inputs, behavioral or structural descriptions and finally the expected outputs. By synthesis, we understand the process of describing the functionality of our VHDL code, using components from different abstraction levels. Thus, in case of the register transfer level, the algorithm is described using arithmetic logic units, registers and finite state machines; in case of the gate-level abstraction, a more in-depth description of the algorithm is generated, using logic gates, flip-flops and parameters such as propagation delays; in case of the lowest abstraction level (i.e. transistor level) the description is done using components such as transistors, resistors and signals represented as voltages.

The methodology of digital design with VHDL starts with *(1) the problem definition* step, where all the inputs and outputs of our system are defined within the port declaration section of our VHDL code (see Fig. 1.21).

Further on, *(2) the behavioral description* of the digital logic intended to implement our algorithm is described using VHDL language constructs. The next step is (3) *the test-bench description*, which delivers stimuli to our digital logic and reads the corresponding outputs in order to verify the functionality of our algorithm implementation. Thus, using the test-bench description the first (4) *simulation of the VHDL* code is performed. Note that the simulation tool takes as input our VHDL code description and generates the register-transfer level components of our design. In case the simulation results are as expected, we proceed to the next step (5) called *synthesis*. There are 3 type of synthesis performed on the VHDL description: a) RTL (Register Transfer Level) synthesis, b) logic synthesis and c) technology mapping. They all generate structural *netlist* files. In case of the first two synthesis steps a) and b), different libraries are used to create structural descriptions of the digital logic using RTL components (e.g. registers, arithmetic logic units) and gate level components (e.g. logic gates, flip-flops), respectively. In case of the third synthesis step, technology dependent libraries (each target device with different libraries) are used to associate components commonly referred as *cells* to the *netlist* description. The resulted netlist is called *cell-level netlist* and is specific for each technology (i.e. ASIC or FPGA device family). At the end of the technology

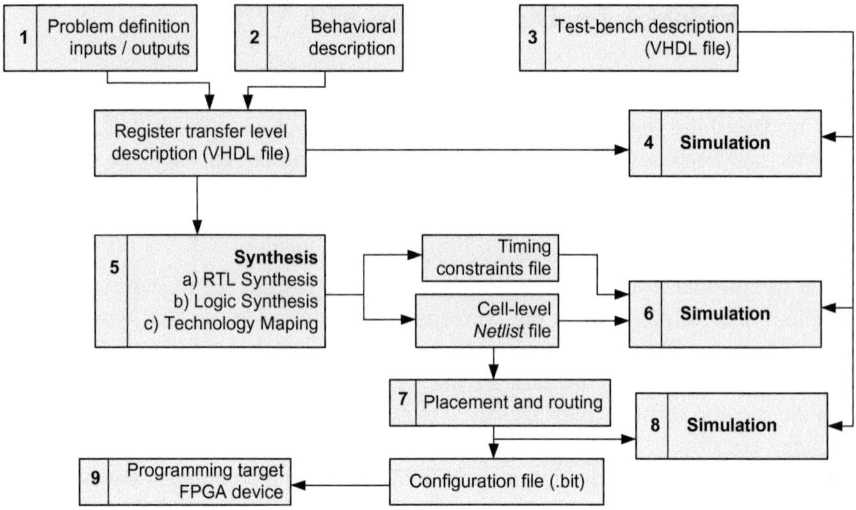

Fig. 1.21 Methodology for digital logic design with VHDL

mapping synthesis step, the timing constraints definitions are also available together with the cell-netlist, which may be used for the post-map simulation procedure (6). In case the simulation results meet the expectations, placement and routing is performed, which generates a file for configuring the target device, step (9). Placement and routing delivers also a timing constraints definition file according to target device specification which may be used for a post place and route simulation to verify the design functionality.

1.9 Conclusions

This chapter summarizes the main language constructs together with the methodology to be used for designing digital circuits and systems with VHDL. Detailed examples are provided for the exemplification of important aspects in digital logic design such as: sequential and combinational logic, structural and behavioral description and finite state machines. Moreover, the benefits of VHDL in testing digital logic by means of test-bench-based simulations are also exploited in case of different hardware architecture such as the architecture for image profile computation. The main differences between the VHDL code for synthesis and the VHDL code for writing test-benches are also underlined. To sum up, this first chapter provides all the necessary information for starting digital logic design with VHDL.

Appendix A

Operators for predefined VHDL data types

Operator	*Data type of* ***a***	*Data type of* ***b***	*Resulted data type*	*Description*
a + b	***integer***	***integer***	***integer***	Addition
a − b				Subtraction
a * b				Multiplication
a / b				Division
a mod b				Modulo
a rem b				Reminder
a ** b				Exponentiation
abs a				Absolute value
not a	***boolean, bit, bit_vector***	***boolean, bit, bit_vector***	***boolean, bit, bit_vector***	Negation
a sll b	***bit_vector***	***integer***	***bit_vector***	Shift-left logical
a srl b				Shift-right logical
a sla b				Shift-left arithmetic
a srl b				Shift-right arithmetic
a rol b				Rotate left
a ror b				Rotate right
a = b	Any type (same of type for both a and b)		boolean	Equal to
a /= b				Not equal to
a < b	Scalar or 1-D array		boolean	Less than
a <= b				Less or equal to
a > b				Greater than
a > = b				Greatenr or equal to
a and b	Boolean, bit, bit_vector (all b, and the result are of the same type)			Logic and
a or b				Logic or
a xor b				Logic exclusive or
a nand b				Logic negative and
a nor b				Logic negative or
a xnor b				Logic exclusive negative or

Appendix B

```
library IEEE;
use IEEE.STD_LOGIC_1164.ALL;
use IEEE.STD_LOGIC_ARITH.ALL;
use IEEE.STD_LOGIC_UNSIGNED.ALL;
entity Comp_Addr is
   PORT
         (Xmax : in std_logic_vector (12 downto 0);
         Ymax : in std_logic_vector (11 downto 0);
         CLK : in  STD_LOGIC;
         Start : in STD_LOGIC;
         Reset: in STD_LOGIC;
         R : in  STD_LOGIC_VECTOR (7 downto 0);
         G : in  STD_LOGIC_VECTOR (7 downto 0);
         B : in  STD_LOGIC_VECTOR (7 downto 0);
         DataRAM_X : out std_logic_vector (31 downto 0);
         DataRAM_Y : out std_logic_vector (31 downto 0);
         X : out std_logic_vector (12 downto 0);
         Y : out std_logic_vector (11 downto 0);
         done : out STD_LOGIC;
         DBG_we : out std_logic;
         DBG_sum_x : out std_logic_vector (31 downto 0)
         );
end Comp_Addr;
architecture Behavioral of Comp_Addr is
component x_y
PORT ( Xmax : in std_logic_vector (12 downto 0);
         Ymax : in std_logic_vector (11 downto 0);
         CLK : in  STD_LOGIC;
    Start : in STD_LOGIC;
         Reset: in STD_LOGIC;
         X : out STD_LOGIC_VECTOR (12 downto 0);
    Y : out STD_LOGIC_VECTOR (11 downto 0);
         done: out STD_LOGIC
    );
end component;
 COMPONENT RGBtoY
   PORT(
      CLK : IN  std_logic;
      START : IN  std_logic;
      R : IN  std_logic_vector(7 downto 0);
      G : IN  std_logic_vector(7 downto 0);
      B : IN  std_logic_vector(7 downto 0);
```

```
      Y : OUT  std_logic_vector(7 downto 0);
      Done : OUT  std_logic
     );
   END COMPONENT;
component AdderX
         PORT (   a_lum : in std_logic_vector(7 downto 0);
                b_ram : in std_logic_vector(31 downto 0);
                sum_x : out std_logic_vector (31 downto 0)
                );
end component;
component AdderYY
         PORT (  a_lum : in std_logic_vector(7 downto 0);
                b_ram : in std_logic_vector(31 downto 0);
                sum_Y : out std_logic_vector (31 downto 0)
                );
end component;
component RAM_X
 Port (CLK : in  STD_LOGIC;
   we : in std_logic;
    addr : in std_logic_vector(12 downto 0);
    data : in std_logic_vector (31 downto 0);
    data_out : out std_logic_vector (31 downto 0)
         );
end component;
component RAM_Y
 Port (CLK : in  STD_LOGIC;
      we : in std_logic;
      addr : in std_logic_vector(11 downto 0);
      data : in std_logic_vector (31 downto 0);
      data_out : out std_logic_vector (31 downto 0)
      );
end component;
 signal done_Y_Lum : std_logic;
 signal Y_Lum: std_logic_vector(7 downto 0);
 signal log_done : STD_LOGIC;
 signal data_in : std_logic_vector (31 downto 0);
 signal data_in2 : std_logic_vector (31 downto 0);
 signal sum_x : std_logic_vector (31 downto 0): =  (others => '0');
 signal sum_y : std_logic_vector (31 downto 0): =  (others => '0');
 signal data_out_mem : std_logic_vector (31 downto 0);
 signal data_out_mem2 : std_logic_vector (31 downto 0);
 signal X_copy,X_addr : STD_LOGIC_VECTOR (12 downto 0);
 signal Y_copy, Y_addr : STD_LOGIC_VECTOR (11 downto 0);
 signal done_pipe: std_logic_vector (1 downto 0): = (others  => '0');
 signal read_done_intern : std_logic;
```

```
 signal read_done_intern2 : std_logic;
 signal we_in : std_logic;
 signal we_in2 : std_logic;
 signal res : std_logic: = '0';
 signal init_0 : std_logic_vector (31 downto 0): = (others => '0');
begin
 uut: RGBtoY PORT MAP (
      CLK => CLK,
      START => START,
      R => R,
      G => G,
      B => B,
      Y => Y_Lum,
      Done => done_Y_Lum
     );
  uut3: x_y PORT MAP (
      CLK => CLK,
      Xmax => Xmax,
      Ymax => Ymax,
      Reset => Reset,
      start => done_Y_Lum,
      X => x_copy,
      Y => y_copy,
      done => Res
     );
uu4: AdderX Port map ( a_lum => Y_Lum,
          b_ram => data_out_mem,
          sum_x => sum_x
          );
uu7: AdderYY Port map ( a_lum => Y_Lum,
             b_ram => data_out_mem2,
             sum_y => sum_y
             );
uut5: RAM_X Port map (CLK => CLK,
             we => we_in,
             addr => x_copy,
             data => data_in,
             data_out => data_out_mem
             );
uut6: RAM_Y Port map (   CLK => CLK,       -done
             we => we_in2,       -done
             addr => y_copy,        - done
             data => data_in2,
             data_out => data_out_mem2                          );
X <= x_copy;
Y <= y_copy;
```

```
we_in <= res;
we_in2 <= res;
data_in <= sum_x;
data_in2 <= sum_y;
done <= res;
DBG_we <= we_in;
DBG_sum_x <= sum_x;
DataRAM_X <= data_out_mem;
DataRAM_Y <= data_out_mem2;
end Behavioral;
```

Appendix C

```
————————— FSM for memory controller ——————————————————
library IEEE;
use IEEE.STD_LOGIC_1164.ALL;
use IEEE.STD_LOGIC_ARITH.ALL;
use IEEE.STD_LOGIC_UNSIGNED.ALL;
entity mem_contr is
port(   read_write: in std_logic;
ready, clk: in std_logic;
oe, we: out std_logic);
end mem_contr;
architecture FSM of mem_contr is
type tip_stare is (idle, decision, read, write);
signal STATE, NEXT_STATE: tip_stare;
begin
proc_comb: process (STATE, read_write, ready)
begin
  case STATE is
when idle => oe <= '0'; we <= '0';
  if (ready = '1') then
            NEXT_STATE <= decision;
  else NEXT_STATE <= idle;
  end if;
when decision => oe <= '0'; we <= '0';
  if (read_write = '1') then
NEXT_STATE <= read;
  else - read_write = '0'
NEXT_STATE <= write;
  end if;
```

```
when read => oe <= '1'; we <= '0';
  if (ready = '1') then
NEXT_STATE <= idle;
  else - ready = '0'
NEXT_STATE <= read;
  end if;
when write => oe <= '0'; we <= '1';
  if (ready = '1') then
NEXT_STATE <= idle;
  else - ready = '0'
NEXT_STATE <= write;
  end if;
end case;
end process proc_comb;
proc_secv: process (clk)
  begin
if reset = '1' then
  STATE <= IDLE;
else
  if (clk'event and clk = '1') then
    STATE <= NEXT_STATE;
  end if;
end if;
end process proc_secv;
end FSM;
————————— end of FSM for memory controller ——————
```

References

1. P.P. Chu, *Introduction to Digital System Design—Chapter 1 from RTL Hardware Design Using VHDL: Coding for efficiency, Portability and Scalability* (Wiley, Cleveland State University, Ohio, (2006)
2. D.L. Perry, *VHDL Programming by Example*, Fourth Edition (McGraw-Hill, New York), (2002)
3. V.A. Pedroni, *Circuit Design with VHDL* (MIT Press Cambridge, Massachusetts, London, England), (2004)

Chapter 2
Hardware Architectures for Channel Encoding in Information Transmission Systems

Nowadays, digital representation of any kind of information source is common. Speech waveforms, image waveforms, text files and any other information sources are represented as binary sequences. In order to pass the information from source to destination, the source output, represented as a binary sequence, is further on converted into a form suitable for transmission over a particular physical media (e.g. cable, optical fiber etc.). In other words, information transmission systems are used to pass digital sequences from source to destination trough a particular physical media defined as the communication channel. Considering the aforementioned principle of digital communications, two fundamental ideas are distinguished: communication sources are viewed as digital sequences and communication channel demand the encoding of the digital sequences in a form suitable for reliable transmission [1]. Consequently, digital representation of the information sources is known as *source coding*, whereas the conversion of the source digital sequences in a form suitable for the communication channel is called *channel encoding*. The present chapter presents introductory sections for both information transmission systems and channel encoding, followed by hardware implementations of coder and decoder architectures in case of linear block codes (i.e. Hamming and cyclic codes).

2.1 Introduction to Information Transmission System

Accurate and timely data, presented in a context that gives it meaning and relevance can be defined as *information* [2]. The word information comes from ancient Greek, and is derived from the words "eidos" (idea) and "morphe" (shape, form). Joining the two terms suggests that, the mind interpretation of objects generates information.

With regards to the *Information and Communication Technologies* (ICT), information is embedded into a physical form (e.g. electromagnetic wave) in order

B. Belean, *Application-Specific Hardware Architecture Design with VHDL*, Signals and Communication Technology,
https://doi.org/10.1007/978-3-319-65025-8_2

to be transmitted through a communication channel. Consequently, in ICT information involves an information source *S*, a destination *D* and a physical medium (i.e. transmission channel) that assures the information transmission from *S* to *D*. The information source can be discrete (digital source), or continuous (signal source). In our case, 0 and 1 symbols are used for information representation, thus information is discrete and specific for digital communications.

In digital communication, the ensemble of interdependent blocks used to transfer the information from source to destination is called *Information Transmission System* (ITS). As mentioned before, source coding and channel coding are independently performed using a binary interface between source and channel. Using digital sequences for information transmission is specific for digital communications systems.

The main advantages of digital communication systems are, on one hand, the possibility of flexible implementation and low cost considering the advent of digital logic circuits, and, on the other hand, the reliability and the possibility to ensure information confidentiality [2]. Digital communications come with the disadvantage, the increased bandwidth compared to analogue information transmission systems. Nevertheless, the disadvantage is diminished due to the existing computational power and the possibility of data compression.

2.1.1 Modelling an Information Transmission System

The schematic diagram depicted in Fig. 2.1 represent a digital ITS.

The *source encoder* converts the input signals into a sequence of bits *i*. For example, a number of *m* source messages to be transmitted are coded using *n* bits for each source message, where $m < 2^n - 1$. The main benefits of this source message conversion into a digital sequence are the inexpensive digital hardware and possibility of source/channel separation.

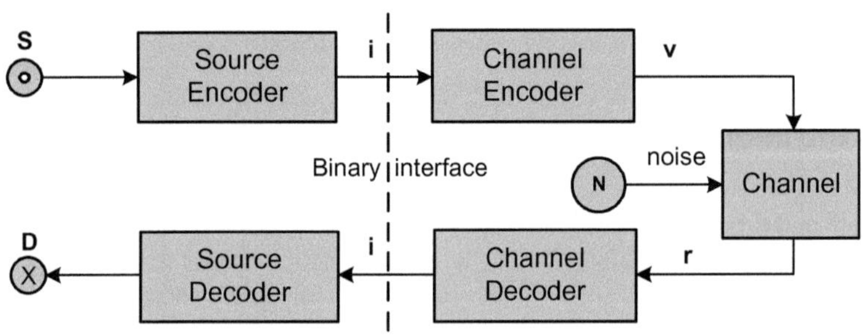

Fig. 2.1 Schematic model of an ITC for information transmission from sours S to destination D. The source messages are encoded in binary sequences i, and further on converted into a codeword *v* for channel transmission. The channel decoder performs error detection and correction on the received codeword v

The transmission channels, especially the electromagnetic field specific for wireless communication, are susceptible to noise which may alter the signals. The role of the channel encoder and decoder is presented next, considering noisy communication channels. To start with, the noise in communication channel is shortly described. Commonly, a channel model includes an input waveform *x(t)*, a noise waveform *n(t)* and the output waveform *y(t)*. In our case, the noise waveform is considered an additive white Gauss noise (AWGN). The input signal is characterized by its power *P*, whereas the communication channel is of bandwidth *B*. A characteristic of the communication channel is the signal per noise ratio SNR, given by the input signal power level compared to the background noise power, as denoted by ξ from Eq. (2.1). The SNR is often expressed in dB.

$$\xi = \frac{\overline{[y(t)]^2}}{\overline{[n(t)]^2}}, \xi_{[dB]} = 10 \log_{10} \xi \tag{2.1}$$

Considering channel noise levels, Shannon defined the maximum decision rate D_{max} [bits/s] that can be error free transmitted through an AWGN communication channel of bandwidth *B*. By decision rate we understand the number of bits per second delivered by the channel encoder block. Consequently, the channel capacity C is given by Eq. (2.2).

$$C = B \log_2(1 + \frac{P}{N}) = B \log_2(1 + \frac{P}{BN_0}) \tag{2.2}$$

where *N* is the noise power and N_0 is the noise power per unit bandwidth. The channel capacity as previously defined represents a theoretical limit impossible to be reached in real transmission systems. Nevertheless, state-of-the-art channel coding schemes together with remarkable computational power offered by digital hardware such as GPU or FPGA allow to closely approach this channel capacity, whereas high-throughput is also delivered. A classic channel coding scheme is detailed further on.

The channel coding, i.e., the *channel encoder* and the *channel decoder*, converts the information *i* into a codeword *v*, which is transmitted to the channel. At destination, the channel decoder block receives the codeword *r*, and decodes the sequence of information bits *i'*. These conversions are performed in order to ensure a high degree of accuracy at destination, in spite of the noisy transmission channels. In other words, redundant bits are added by the channel encoder to the information bits *i* and the codeword *v* is obtained. Based on the added redundant bits, the decoder performs error detection and correction of the information bits. A more detail description of error control using channel encoding is presented in the next section.

2.2 Introduction to Channel Encoding for Error Control

When information transmission is performed through a noisy communication channel, the signal carrying the information is susceptible to alterations due to the channel noise. Nowadays, considering the high impact of communication in economic and social life, data transmission requires increased reliability, high speed and increased throughput. Thus, protection against channel noise is considered in order to eliminate this unwanted effect. The error protection is performed through channel coding. The history of error control coding starts in mid 20th century and began with Hamming codes and reached to more powerful error correction codes such as low density parity check codes (LDPC) or turbo-codes, trying to limit technically errors effect in applications. The solution for achieving error protection while transmitting information was proposed by C. E. Shannon is known as Shannon second theorem namely noisy channels coding theorem [3].

Basically, the theorem proves that on a noisy channel having the capacity *C*, a real time transmission with *D* bit rate and an error probability *P(E)* as small as wanted, using uniform codes of length *n*, such that: $\mathrm{P(E)} \leq 2^{-\mathrm{n \cdot e(D)}}$ (*e(D)* is a positive function of *D*, completely determined by channel characteristics). This means an error probability *P(E)* however small can be obtained in two ways: by increasing *D* which lead to inefficient channel usage and by using large codewords by adding redundant bits [2]. The last approach is used in practice for error control codes for data transmission over noisy channels. The strategies for error control in data transmission are classified as follows: (i) error detection which informs the transmitter about the need for retransmission of erroneous data (ARQ—automatic repeat request); (ii) forward error correction (FEC) in which case the channel coding approach automatically corrects an amount of errors at the receiver side; (iii) error correction and detection which involves error correction through coding according to error correction code capacity and repeat transmission for the rest of erroneous combinations (e.g. radio transmissions). To conclude, error detection and correction is achieved by channel encoding.

2.2.1 *Representation of Error Control Codes*

The representations used for binary code sequences can be classified as matrix, vector, polynomial and geometrical representations described as follows:

- matrix representation implies all the code words are stored in matrix, excepting the zero components; let *m* be the number of codewords and *n* be the length of one codeword, than the whole code may be stored in the next matrix ($a_{ij} \in \{0; 1\}$):

$$\mathbf{C} = \begin{bmatrix} a_{11} & a_{12} & \dots & a_{1n} \\ a_{21} & a_{22} & \dots & a_{2n} \\ \dots & \dots & \dots & \dots \\ a_{m1} & a_{m2} & \dots & a_{mn} \end{bmatrix} \begin{matrix} \Rightarrow \mathbf{v_1} \\ \Rightarrow \mathbf{v_2} \\ \dots \\ \Rightarrow \mathbf{v}_m \end{matrix} \quad (2.3)$$

- vector representation of codes considers each code word of length n as a n—length vector from the vector space $\boldsymbol{V_n}$. In other words each sequence of length n is represented by a vector as shown below:

$$\mathbf{v} = (\, a_1 \quad a_2 \quad \dots \quad a_n \,), \quad (2.4)$$

where $a_i \in \{0; 1\}$ for binary codes.
- polynomial representation of a length n codeword $\boldsymbol{v}$ is made through a polynomial of degree n and unknown x, for which the coefficients are represented by the codeword bits a_i. Polynomial representation example:

$$v(x) = a_{n-1}x^{n-1} + a_{n-2}x^{n-2} + \; ... + a_1x + a_0 \quad (2.5)$$

- geometric representation implies that the code words of length n are represented as points which defines the peaks of a geometrical shape, in a n-dimensional space; the main benefit is that the representation allows using a series of well known proprieties of geometric figures to the codes.

Within the current chapter, vector representation and polynomial representation of codes will be used in case of Hamming codes and cyclic codes respectively.

2.2.2 *Classification of Error Control Codes*

Considering the nature of the processed information, the error control codes used to detect, to correct or for both error correction and detection can be classified as:

- block codes in which case, the information at the encoder input is divided into blocks of m symbols/bits to which the encoder adds k control symbols resulting in an $n = m + k$ length codeword.
- convolutional codes in which case the information bits are processed in a continuous way; e.g. an n bits codeword is coded into another n bits codeword based on the coding relations.

Taking into account the types of the error propagated through the communication channel, the error control codes are:

- codes for independents error correction;
- codes for burst error correction;

Based on the possibility of having or not, the information and control symbols positions well defined in a codeword, the error control codes can be classified in:

- systematic codes for which both the information bits i and the control bits c positions are clearly defined;
- non-systematic codes in which case the information bits and control bits are not given in clear considering the codeword structure.

2.2.3 Error Control Codes Parameters

The performances and characteristics of error control codes are illustrated by a set of parameters which are further on detailed. Let us consider the codeword length given by parameter n and the number of information bits is m and the control bits k in case the information bits and control bits are separable.

Redundancy (R), as shown in Shannon second theorem, is used to achieve error protection by adding supplementary bits for error detection or correction. Depending on the type of error control codes (i.e. separable or non-separable corresponding to the possibility to separate the redundant bits from information bits or not) the redundancy is computed as presented bellow.

For separable codes, redundancy is given by:

$$R = \frac{k}{n}, \tag{2.6}$$

whereas for non-separable codes,

$$R_r = \frac{\log_2(\text{total no. of nlength sequences}) - \log_2(\text{no. of codewords})}{log_2(\text{total number of nlength sequences})} \tag{2.7}$$

Detection/correction capacity (C_d/C_s) represents the ration between the number of detectable/correctable erroneous sequences and the total number of erroneous sequences.

Code distance (d) is a parameter indicating the capacity of the code to detect and correct errors. The relation between the code distance and the error detection/correction capability are demonstrated in [4].

Let us first define the Hamming weight w of a codeword $\boldsymbol{v}$ by the number of non-zero symbols of the respective codeword. Moreover, let $\boldsymbol{v}_i$ and $\boldsymbol{v}_j$ be the two codewords for which the hamming distance is intended to be computed. The $d(\boldsymbol{v}_i, \boldsymbol{v}_j)$ is defined by the number of positions for which $\boldsymbol{v}_i$ and $\boldsymbol{v}_j$ differ, and denotes the hamming distance: $d(\mathbf{v}_i, \mathbf{v}_j) =: \sum_{k=1}^{n} a_{ki} \oplus a_{kj}$ where $\mathbf{v}_i = (a_{1i}, a_{2i}, \ldots, a_{ni})$, $\mathbf{v}_j = (a_{1j}, a_{2j}, \ldots, a_{nj})$ and $\oplus$ is modulo 2 summation.

As referred to the error correction capability related to the hamming distance, the necessary and sufficient condition for a code to correct maximum t errors is $d \geq 2t + 1$. Considering the error detection capacity, the necessary and sufficient condition for a coded to detect maximum *e* errors is $d \geq e + 1$. The necessary and sufficient condition for a code to simultaneously correct maximum t errors and detect maximum *e* errors is $d \geq t + e + 1, \; e > t$ [2].

2.3 Block Codes

Considering a block code, information from a binary source is divided into *m*-bits blocks representing the information bits ***i***. For each information symbols *m* we add *k* control symbols according to a coding rule. The result is a codeword ***v*** of length *n*, according to Eq. (2.8).

$$n = m + k \tag{2.8}$$

Considering such a structure, the number of resulted codewords is given by M, with $M = 2^m$. Let $\mathbf{V_n}$ (dimension n) denote the space of vectors containing all the codewords of *n* bits (with coefficients in Galois Field GF(2)). In case 2^m codewords form a vector space of dimension *m* which is a subspace *C* of the space $\boldsymbol{V_n}$, we call the block code linear, and the linear block code is denoted by the pair *(n, m)*. Such linear structures are very important for the practical applications for the block codes. Another interpretation of the block codes linearity is that, a block code is linear if and only if the modulo 2 sum of two code words is also a code word. It follows that the vector space of dimension *n*, $\boldsymbol{V_n}$ containing the set of distinct combinations that can be generated with *n* bits (2^n) is divided into two distinct sets:

C—the set of code-words,

$F = V_n - C$—the set of false code-words.

The linear block codes were invented after Shannon gave his second theorem (1948). Thus, R. Hamming and Golay introduced the systematic liner codes (1950), whereas the unified theory for linear codes was documented in 1960. Many practical applications in information theory and coding were developed since, which make use of linear block codes [5]. Further on, Hamming codes and cyclic codes are exemplified, but first the coding equations are detailed.

2.3.1 Coding Equations

As mentioned before, the linear code *C(n,m)* is an *m*-dimensional subspace of V_n. Consequently, the entire set of codewords can be generated by the linear combinations of the *m* linear independent vectors belonging to *C*. Let the set of linear independent code vectors be denoted by $\mathbf{g_i}, \; i = \overline{1, m}$. This set of vectors can be written as the matrix from Eq. (2.9) and represents the *code generating matrix* $G_{[mxn]}$.

$$\mathbf{G} = \begin{bmatrix} g_{11} & g_{12} & \cdots & g_{1n} \\ g_{21} & g_{22} & \cdots & g_{2n} \\ \cdots & \cdots & \cdots & \cdots \\ g_{m1} & g_{m2} & \cdots & g_{mn} \end{bmatrix} \begin{matrix} \Rightarrow \\ \Rightarrow \\ \cdots \\ \Rightarrow \end{matrix} \begin{bmatrix} \mathbf{g_1} \\ \mathbf{g_2} \\ \\ \mathbf{g_m} \end{bmatrix} \tag{2.9}$$

where g_{ij} are binary symbols.

The encoding equation is given by:

$$\mathbf{v} = \mathbf{iG} = [i_1 i_2 \ldots i_m] \begin{bmatrix} \mathbf{g_1} \\ \mathbf{g_2} \\ \\ \mathbf{g_m} \end{bmatrix} = \mathbf{i_1 g_1} + \mathbf{i_2 g_2} + \ldots + \mathbf{i_n g_n} \tag{2.10}$$

In order to obtain a systematic structure of the codeword **v**, denoted by $\mathbf{v}' = [\mathbf{i} \quad \mathbf{c}]$ the generating matrix **G** must have the *canonical form*:

$$\mathbf{G}' = [\mathbf{I_m P}] \tag{2.11}$$

where $\mathbf{I_m}$, is the unit matrix of order *m*.

The *m* information symbols are found unchanged in the codeword **v** structure and that the *k* control symbols are linear combinations of information symbols. Let c_i be the *k* control symbols and i_j the *m* information symbols. The control symbols are computed based on the information bits, according to the f_i functions as shown by Eq. (2.12).

$$c_i = f_i(i_j), \; i = \overline{1,k}, \; j = \overline{1,m} \tag{2.12}$$

Equation (2.12) are known as the parity check equations. Considering the binary vector spaces properties [6], if we have a space *C* of dimension *m*, then always exists an orthogonal space C* included in the vector space C such that a codeword $v^{\in}C$ is orthogonal in C*. The linear independent k vectors belonging to the orthogonal space C* can be put in a matrix $\mathbf{H_{[kxn]}}$ named the parity check matrix.

The spaces **C** and $\mathbf{C}^*$ being two orthogonal spaces means that the two matrices **G** and **H** are also orthogonal, as expressed by Eq. (2.13).

$$\mathbf{GH^T} = \mathbf{HG^T} = \mathbf{0} \tag{2.13}$$

The coding Eq. (2.10), relative to the vector space $\mathbf{C}^*$ becomes:

$$\mathbf{Hv^T} = \mathbf{0} \tag{2.14}$$

2.3.2 Decoding Equations

Let us consider the communication channel and from Fig. 2.2. The information bits $\boldsymbol{i}$ are coded using the channel coder and the codeword v is obtained which is transmitted through the communication channel. The binary codeword $\boldsymbol{v}$ is affected by channel noise, and, consequently the error vector $\boldsymbol{e}$ contains "1" values as additive noise for the bits position. The $\boldsymbol{e}$ vector is unknown at the receiver side. Thus, the vector $\boldsymbol{r}$ is received which contains errors.

The received codeword represented by the binary vector $\boldsymbol{r}$ is given by Eq. (2.15).

$$\mathbf{r} = \mathbf{v} + \mathbf{e} \tag{2.15}$$

The error vector $\boldsymbol{e}$ indicates the additive errors due to the noise within the communication channel. The sign + is the modulo 2 summation of the two vectors, codeword $\boldsymbol{v}$ and the error vector $\boldsymbol{e}$, respectively.

Considering the matrix representation, the error vector e may be written as:

$$\mathbf{e} = \begin{bmatrix} \mathrm{e}_1 & \mathrm{e}_2 & \ldots & \mathrm{e}_\mathrm{n} \end{bmatrix}$$

where an e_i value of "1" means an error at position I, whereas a value of "0" means an correct bit within the received codeword r. As far as for the *decoding equation* at the receiver side, it is based on the syndrome S, as denoted by Eq. (2.16). The syndrome represents a column matrix with k elements.

$$\mathbf{H}\mathbf{r}^\mathrm{T} = \mathbf{S} \tag{2.16}$$

By replacing **r** vector we obtain:

$$\mathbf{H}(\mathbf{v} + \mathbf{e})^\mathrm{T} = \mathbf{H}\mathbf{e}^\mathrm{T} = \mathbf{S} \tag{2.17}$$

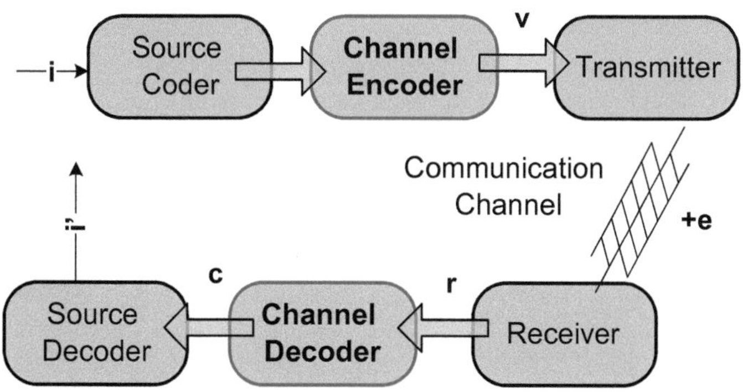

Fig. 2.2 Channel coder/decoder implementation scheme; blocks of $\boldsymbol{i}$ information bits are coded and transmitted through the communication channel; the receiver gets the codeword transmitted to the channel ($\boldsymbol{r}$) and corrects the errors denoted by $\boldsymbol{e}$

Equation (2.17) denotes that the syndrome S depends only on the channel errors. Consequently, if there are no errors or the errors cannot be detected, the syndrome S = 0. Meanwhile, if the syndrome $\mathbf{S} \neq \mathbf{0}$ then the errors are detected. In case the error correction is wanted, it is necessary to determine the error position based on the resulted syndrome. This is the case for the Hamming codes. An example of a coder/decoder implementation for the Hamming codes is detailed further on.

2.4 Hamming Coder/Decoder Implementations

Hamming codes were introduced after Shannon second theorem by R. Hamming in 1950. In our case, we detail a Hamming group code for the correction of single errors.

The characteristics of this code are:

- the code length can be determined by:

$$n = 2^k - 1 \tag{2.18}$$

- the codeword structure is given by the Eq. (2.19), where a_i represent the information symbols and c_i represent the control (parity check) symbols;

$$\mathbf{v} = \left[\mathrm{c_1 c_2 a_3 c_4 a_5 a_6 a_7 c_8 a_9 \ldots a_n}\right] \tag{2.19}$$

- the control symbols are placed at positions 2^{i} within the codeword, with $\mathrm{i} = \overline{0, \mathrm{k} - 1}$;
- the control matrix **H** is given by Eq. (2.20), where each column $\boldsymbol{h_i}$ expresses in binary natural code (BN) its position with the less significant bit LSB on the kth line;

$$\mathbf{H}_{[\mathrm{k} \times \mathrm{n}]} = \left[\mathbf{h_1} \quad \ldots \quad \mathbf{h_i} \quad \ldots \quad \mathbf{h_n}\right] \tag{2.20}$$

The coding relationships are determined using Eq. (2.14). Consequently, the control symbols are expressed as a linear combination of information symbols, as expressed by Eq. (2.12).

Regarding the decoding process, having the received codeword $\boldsymbol{r}$, it verifies the decoding Eq. (2.16). Thus the **S** syndrome is determined by Eq. (2.21).

$$\mathbf{S} = \left[\mathbf{h_1} \quad \ldots \quad \mathbf{h_n}\right] \begin{bmatrix} 0 \\ \ldots \\ \mathrm{e_i} \\ \ldots \\ 0 \end{bmatrix} = \mathbf{h_i} \tag{2.21}$$

When only one error occurs, the syndrome indicates a binary number h_i, corresponding to the error position within the codeword $\boldsymbol{r}$. Thus, using a binary-decimal conversion, we can determine from the S syndrome the erroneous position. In case more than one error occurs, a major disadvantage comes up. The supplementary errors are not detected, whereas the decoder introduces supplementary errors. To deal with the aforementioned disadvantage, modified Hamming codes are available (extended or shortened Hamming codes) which allow supplementary errors detection or correction.

Following the aforementioned descriptions for coding and decoding processes, hardware architectures are built both for the coder and the decoder, in case of a Hamming (7, 4) group code.

2.4.1 Encoder Implementation

The Hamming (7, 4) codeword structure is presented as follows:

$$\mathbf{v} = [c_1 \ c_2 \ a_3 \ c_4 \ a_5 \ a_6 \ a_7] \tag{2.22}$$

The parity check matrix **H** and the encoding equations are denoted by the Eq. (2.23).

$$\mathbf{H} = \begin{bmatrix} 0 & 0 & 0 & 1 & 1 & 1 & 1 \\ 0 & 1 & 1 & 0 & 0 & 1 & 1 \\ 1 & 0 & 1 & 0 & 1 & 0 & 1 \end{bmatrix} \qquad \mathbf{H} \cdot \mathbf{v}^T = \mathbf{0} \tag{2.23}$$

The control symbols are expressed as a linear combination of information symbols according to the following set of equations:

$$\begin{aligned} c_1 &= a_3 + a_5 + a_7 \\ c_2 &= a_3 + a_6 + a_7 \\ c_4 &= a_5 + a_6 + a_7 \end{aligned} \tag{2.24}$$

Example 2.1 Hardware architecture for the Hamming encoder

The hardware architecture corresponding to the Hamming encoder is described in Fig. 2.3a. The main components are: a shift register with parallel load for each codeword; three adders for calculating the parity check bits c_1, c_2 and c_4. According to the equations set (21), the control bits are calculated and loaded into the shift register *RD* in the corresponding positions: *1*, *2* and *4*. In this manner, the code word $\boldsymbol{v}$ is formed using the shift register *RD*.

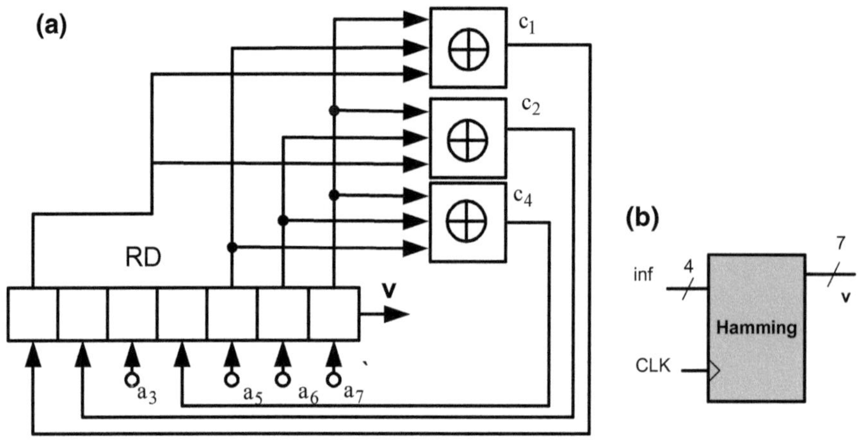

Fig. 2.3 **a** Hamming coder implementation using shift register RD, **b** logic block for the Hamming encoder (inputs/outputs)

The logic block for the Hamming encoder intuitively called Hamming is illustrated in Fig. 2.3b. The Hamming logic block has a 4 bits input representing the information bits and an output of 7 bits for the codeword resulted after encoding. The corresponding VHDL code for the entity associated with the Hamming encoder is presented next:

```
——————— Entity declaration———————————————————————————
entity Hamming is
Port ( Clk: IN std_logic;
       inf: IN std_logic_vector(3 downto 0);
       data_paralel:   OUT std_logic_vector (6 downto 0));
END Hamming;
——————— End of entity declaration————————————————————
```

The corresponding VHDL code which describes the Hamming encoder functionality is presented next. A shift register *RD* with parallel load is used to store the received codeword. The register is represented by the *temp* internal signal, whereas the parallel load operation is performed through the concurrent statement *temp (j) = inf(i)*. Note that *j = {6, 5 4, 2}*, the position of information bits, whereas *i = 0 to 3*. The 7 bits output is also assigned in a concurrent manner to the *temp* register output (e.g. *data_parallel* <= *temp*). In order for the *temp* signal to represent a register within the behavioral architecture, the control bits are computed by *xor* operator and assigned on the rising edge of the *clk* input, on the positions 0, 1, and 3 (e.g. `temp(0) <=inf(3) xor inf(0) xor inf(1);`).

```
——————— Behavioral description of the Hamming encoder ———————
architecture Behavioral of Hamming is
```

```
signal temp: std_logic_vector (6 downto 0);
begin
    process (clk)
    begin
        if clk'event and clk = '1' then
            temp(3) <=inf(1) xor inf(2) xor inf(3);
            temp(1) <=inf(0) xor inf(2) xor inf(3);
            temp(0) <=inf(3) xor inf(0) xor inf(1);
        end if;
    end process;
  data_paralel <= temp;
  temp(6) <=inf(3);   temp(5) <=inf(2);
  temp(4) <=inf(1);   temp(2) <=inf(0);
end Behavioral;
————————End of behavioral description of the Hamming encoder ————
```

Figure 2.4 illustrates the parallel load of the input (i.e. the information bits *inf*) into the shift register RD at simulation time t_0 and t_2. The resulted codewords after adding the control bits are underlined at simulation time t_1 and t_3 as underlined in Fig. 2.4.

Example 2.2—Hardware architecture for the Hamming decoder
The Hamming decoder receives the codeword r through the communication channel. In case of the Hamming (7, 4) code, the error prone received codeword is denoted by Eq. (2.25).

$$r = [r_1 r_2 r_3 r_4 r_5 r_6 r_7] \tag{2.25}$$

Taking into account Eq. (2.14), the Hamming decoder corrects or detects the transmission errors based on the syndrome S. The following cases are distinguished based on the S syndrome values:

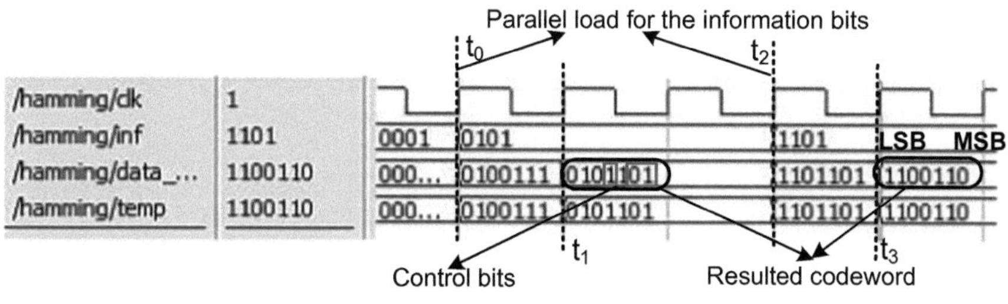

Fig. 2.4 **a** Simulation of the Hamming encoder; the resulted codewords corresponding to the information bits *bx0101* and *bx1101*are underlined

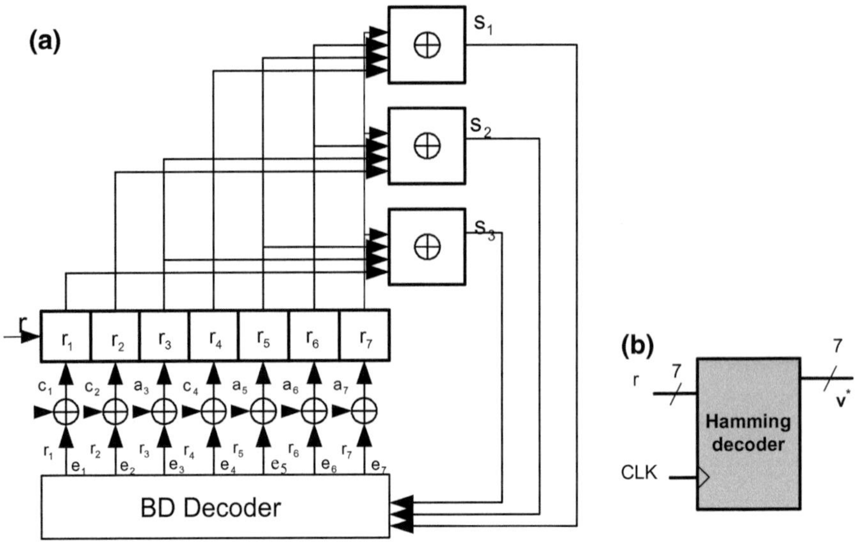

Fig. 2.5 **a** Block diagram for the Hamming decoder composed of a shift register, adders and binary to decimal decoder (BD decoder), **b** the black-bocx corresponding to the Hamming decoder

- $S = 0$ meaning there are no errors or the errors are not detected;
- $S \neq 0$ meaning the error is found at the position given by the binary representation of the syndrome S (e.g. the syndrome $S = [011]$ denotes an error on the 3rd position within the received codeword r).

The hardware architecture for the Hamming decoder based on syndrome decoding is described in Fig. 2.5a. The main components are: a shift register where the received codeword is loaded in parallel, three adders for computing the syndrome bits ($S = [s_1, s_2, s_3]$), a binary to decimal decoder (i.e. BD decoder) for computing the position of the erroneous bit. Once the received codeword r is loaded in the shift register, the syndrome S is computed. Further on, the computed binary value of the S syndrome is decoded into a 7 bits binary vector e, which marks the position of the erroneous bit. The received codeword r is corrected by adding "1" logic value to the position specified by the BD decoder. Thus, the corrected codeword is denoted by Eq. (2.26).

$$\mathbf{v}^* = \mathbf{r} + \mathbf{e} \tag{2.26}$$

The logic block for the Hamming decoder is illustrated in Fig. 2.5b. It has a 7 bits input *y_in* representing the received codeword through the communication channel and an output of 7 bits (*y_out*) for the corrected codeword resulted after decoding process. There are also the clock signal *clk* and the reset input *reset* available for the Hamming decoder. The corresponding VHDL code for the entity associated with the Hamming decoder is presented next:

```
———— Entity declaration————————————————————
    entity decoder is
    Port ( clk:          in std_logic;
               reset :  in std_logic;
               v_in:    in std_logic_vector(0 to 6);
               v_out:   out std_logic_vector(0 to 6)
               );
      end decoder;
———— End of entity declaration——————————————
```

The functionality of the Hamming decoder is described within its behavioral description detailed in the next VHDL code section. There are two processes describing the functionality of the decoder (lines 5 and 13) together with two concurrent assignments (lines 33 and 34). The signal *temp2* is used within the first process to build a register which stores the syndrome $S = [s_1\ s_2\ s_3]$. The values for the syndrome bits s_1, s_2 and s_3 are computed and stored in *temp2(i)* register cells, according to the code lines 8, 9 and 10. The second sequential process, computes the error vector *err* which specifies the position of the erroneous bit, based on the syndrome *S*. Further on, the concurrent assignments corresponding to the code lines 33 and 34 perform the correction of the codeword *v_in* by adding the error vector *err* to the *temp1* register. It is to be mentioned that, the *temp1* registers store the input codeword *v_in*. Consequently, the last assignment (line 34) delivers the corrected codeword to the decoder output *v_out*.

```
1:   architecture Behavioral of decoder is
2:   signal temp1: std_logic_vector(0 to 6);
3:   signal temp2: std_logic_vector(0 to 2);
4:   signal err: std_logic_vector(0 to 6);


4:   begin
5:   process (clk)
6:           begin
7:           if clk'event and clk = '1' then
8:               temp2(0) <=v_in(3) xor v_in(4) xor v_in(5) xor v_in(6);
9:               temp2(1) <=v_in(1) xor v_in(2) xor v_in(5) xor v_in(6);
10:              temp2(2) <=v_in(6) xor v_in(4) xor v_in(2) xor v_in(0);
11:          end if;
12:    end process;
```

```
    process (clk)
    begin
       if ( clk'event and clk  = '1') then
           if ( reset  =  '1') then
              err <= "0000000";
           else
              case temp2 is
                 when "000"  => err  <= "0000000";
                 when "001"  => err  <= "0000001";
                 when "010"  => err  <= "0000010";
                 when "011"  => err  <= "0000100";
                 when "100"  => err  <= "0001000";
                 when "101"  => err  <= "0010000";
                 when "110"  => err  <= "0100000";
                 when "111"  => err  <= "1000000";
                 when others  => err  <= "0000000";
              end case;
           end if;
       end if;
   end process;
   temp1  <= v_in;
   v_out  <= err XOR temp1;
   end Behavioral;
```

The simulation of the Hamming decoder implementation is presented in Fig. 2.6, where three different situations of the decoding operation mode are underlined. At the simulation time t_0, the decoder input got no errors and consequently, the resulted *err* signal is err = '0000000'. At the simulation time t_1, there is a unique error in the *v_in* codeword, which is corrected using the *err* vector err = '0001000' at the simulation time t_2. The last simulation time t_3 corresponds to an input code word *v_in* having two errors; in this case the decoder introduces an additional error to the output signal *v_out*.

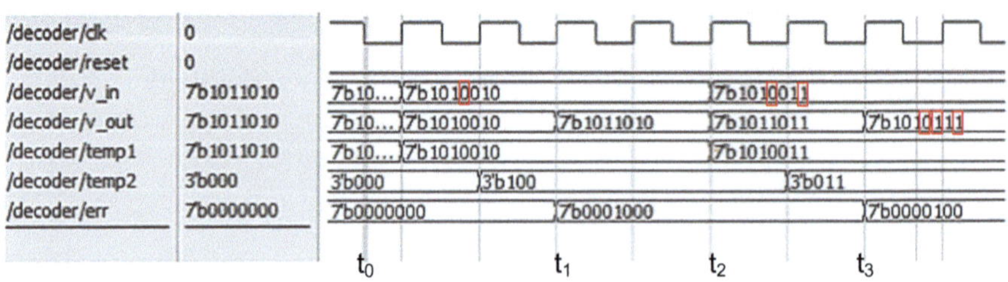

Fig. 2.6 Simulation of the Hamming decoder implementation; the erroneous positions are marked with rectangular areas

2.5 Cyclic Codes Principles

Cyclic codes are a well known subset of the linear block codes commonly used in practice, considering the simplicity of their implementation. The implementation is performed using linear feedback shift registers.

Cyclic codes were first studied by E. Prange in 1957, whereas in 1960 multiple error correction cyclic codes were introduced as BCH codes by R. Rose and P. Chauduri. Moreover, non-binary cyclic codes known as Reed—Solomon codes were proposed in 1960.

Definition

A cyclic code is a linear block code if any cyclic permutation of a codeword is also a codeword. In other words, if a codeword v = (a_0 a_1 … a_n) is included in the set of codewords C, than any cyclic permutation of *v* (i.e. $v^{(1)}$ = (a_n a_0 … a_{n-1}) … $v^{(i)}$ = (a_{n-i} a_{n-i-1} …a_{n-1} a_0 a_1 … a_{n-i+1})) is still a codeword.

Further on, the polynomial representation of the codes is used for describing the cyclic codes coding and decoding processes. Let i = (i_0…i_{m-1}) be the information bits. Thus, the information polynomial *i(x)* of degree *m-1* detailed in Eq. (2.27) describes the information bits.

$$i(x) = i_0 + i_1 x + \ldots + i_{m-1} x^{m-1} \tag{2.27}$$

The codewords of size *n* are chosen as polynomials multiples of a k = n-m degree polynomial known as the code generator polynomial *g(x)* (see Eq. 2.28).

$$g(x) = g_0 + g_1 x + \ldots + g_k x^k, g_k = g_0 = 1 \tag{2.28}$$

Consequently the coding equation is given by:

$$v(x) = i(x)g(x) \tag{2.29}$$

In the current chapter, we will focus on the binary cyclic codes. For binary codes, considering the *c(x)* polynomial of degree *n*, the set of modulo 2 residue classes of c(x) has 2^n elements, out of which, 2^m elements are considered the codeword set that are multiples of the generator polynomial *g(x)*.

The codeword formed with the relation (2.29) leads to a non-systematic codeword structure. In order for the codeword to have a systematic structure, the next steps are followed as presented in [2].

$$\begin{aligned} x^k i(x) &= i_0 x^k + i_1 x^{k+1} + \ldots + i_{m-1} x^{n-1} \\ \frac{x^k i(x)}{g(x)} &= q(x) + \frac{r(x)}{g(x)} \\ v(x) &= x^k i(x) + r(x) = q(x)g(x) \\ &= a_0 + a_1 x + \ldots + a_{k-1} x^{k-1} + a_k x^k + \ldots + a_{n-1} x^{n-1} \end{aligned} \tag{2.30}$$

Thus, a cyclic codeword multiple of *g(x)* is obtained, with the information bits placed on the first m significant positions whereas the control bits are given by the *r (x)* polynomial. The *r(x)* polynomial represents the reminder after the division of $x^k i(x)$ with *g(x)*. In [2] the reader may find how an equation similar with $Hv^T = 0$ is obtained for the coding process. Considering $h(x) = (x^n + 1)/g(x)$, the H matrix is defined as denoted by Eq. (2.31).

$$\mathbf{H}_{[k\times n]} = \begin{bmatrix} 0 & 0 & \cdots & 0 & h_m & h_{m-1} & \cdots & h_1 & h_0 \\ 0 & 0 & \cdots & h_m & h_{m-1} & h_{m-2} & \cdots & h_0 & 0 \\ \vdots & \vdots & \vdots & \vdots & \vdots & \vdots & \vdots & \vdots & \vdots \\ h_m & h_{m-1} & \cdots & h_0 & 0 & 0 & \cdots & 0 & 0 \end{bmatrix} \tag{2.31}$$

To sum up, we present next how the cyclic code size is determined, and how to choose the generator polynomial of degree k in order to correct a number of t errors.

First, the code size is $n = 2^k\text{-}1$, as the cyclic codes are a particular case of linear block codes described in Sect. 3.1.

Secondly, the *g(x)* is chosen as a primitive polynomial of degree *k,* according to the Table 1 from Annex 2. Moreover, a table with generator polynomials for different codeword sizes *n* and errors to be corrected *t* are detailed in Annex 2, Table 2. Further on, digital circuits for cyclic encoding and decoding are presented and implemented using VHDL code.

2.6 Cyclic Codes Encoder and Decoder Implementations

The coding and decoding process for the cyclic codes is performed through the division of $x^k i(x)$ and *r(x)* respectively to the *g(x)* polynomial. *Linear feedback shift registers* with external modulo 2 adders are used for the polynomials division implementation.

The register cells *Cj* connections to the external adders depend on the characteristic of the generator polynomial $g(x) = g_0 + g_1x + \ldots + g_kx^k, \quad g_k = g_0 = 1$. The block scheme is presented in the Fig. 2.7.

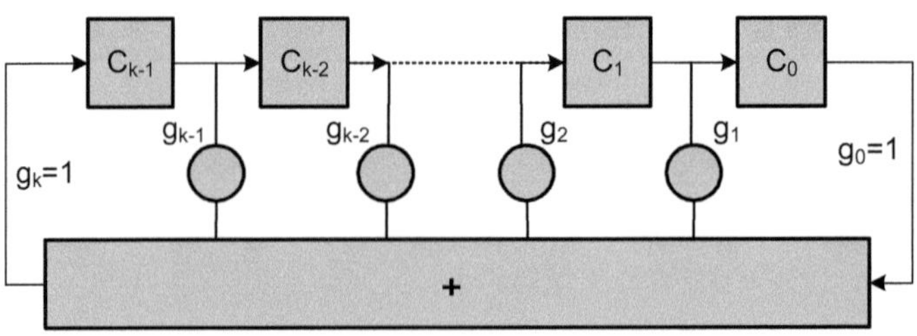

Fig. 2.7 Liner feedback shift registers with external adders for the polynomial division with g(x)

Let the st_iC_j be the state of the register cell C_j at the moment i. Considering a matrix description, the functionality of the linear feedback shift registers is described by Eq. (2.32):

$$\mathbf{S}_i = \mathbf{T}\mathbf{S}_{i-1} \tag{2.32}$$

where

$$\mathbf{S}_i = \begin{bmatrix} st_iC_0 \\ \vdots \\ st_iC_{k-1} \end{bmatrix}, \quad \mathbf{S}_{i-1} = \begin{bmatrix} st_{i-1}C_0 \\ \vdots \\ st_{i-1}C_{k-1} \end{bmatrix}, \text{ and } \quad \mathbf{T} = \begin{bmatrix} 0 & 1 & 0 & \cdots & 0 \\ 0 & 0 & 1 & \cdots & 0 \\ \vdots & \vdots & \vdots & \vdots & \vdots \\ 0 & 0 & 0 & \cdots & 1 \\ g_0 & g_1 & g_2 & \cdots & g_{k-1} \end{bmatrix}. \tag{2.33}$$

The cyclic encoder can be build based on the previously described linear feedback shift register. Thus, another modulo 2 adder S_2 is introduced together with a switch K. This leads to the cyclic encoder from Fig. 2.8.

Concerning the decoder implementation, the information bits i = $[a_{n-1}\ a_{n-2}\ \ldots\ a_{n-m}]$ are delivered to the input sequentially during m clock cycle, whereas the switch K is found at position 1. The encoder output during these first m clock cycles is the same as the input. After m clock cycles, the switch K is in position 2 for the next k clock cycles, while the control bits are computed by dividing $x^k i(x)$ to $g(x)$. At the output we can find along the $m + k$ clock cycles the $v(x)$ polynomial corresponding to the codeword associated with the information symbols from the input (i.e. $v(x) = x^k i(x) + r(x)$). Note that at the end of the n clock cycles, the registers cells are all 0. The Eq. (2.34) expressing the register functionality is illustrated next.

$$\mathbf{S}_i = \mathbf{T}\mathbf{S}_{i-1} + a_i\mathbf{U} \tag{2.34}$$

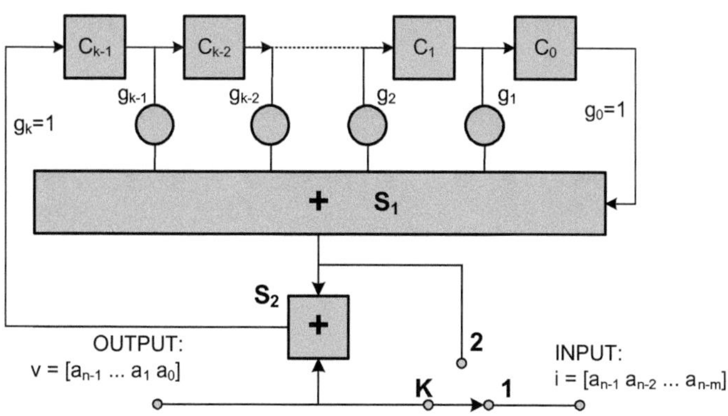

Fig. 2.8 Cyclic encoder with linear feedback shift register and external adders

where a_i is the input signal at the moment i and the U matrix is $U = [0\ 0\ \ldots\ 1]^T$ of size k. Note that during the k clock cycles while the switch K is on position 2, the input is 0 logic. This leads to the value 0 for all the register cells at the end of the n clock cycles (i.e. $S_n = 0$).

The equality $S_n = 0$ and the Eq. (2.31) are used to compute the control symbols based on the information bits (i.e. the encoding equations).

Example 2.3—Hardware architecture for the cyclic encoder implementation We will present how the encoding relation are determined for a cyclic code C(7,4) using the generator polynomial $g(x) = x^3 + x+1$. Further on, the relation will be verified through simulation, using the VHDL implementation of the cyclic encoder having the same size and the same generator polynomial.

The block scheme for the C(7,4) cyclic encoder is detailed in Fig. 2.9.

The encoder operates according to the Table 2.1. Thus, the initial state of the register is $[C_1\ C_2\ C_3] = [0\ 0\ 0]$. For the first $m = 4$ clock cycles, the switch is on position 1, meaning the output is the same as the input (i.e. $i = [a_6\ a_5\ a_4\ a_3]$), whereas the register cells evolve according to the generator polynomial $g(x)$. For the next $k = 3$ clock cycles, the switch is on position 2, meaning the *output* v at

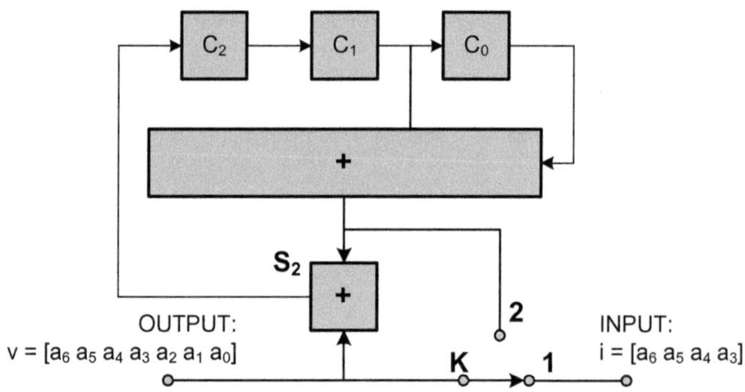

Fig. 2.9 Cyclic encoder for the C(7,4) cyclc code with $g(x) = x^3 + x + 1$

Table 2.1 Hardware resource usage for the implementation of 10 hardware architectures for Canny edge detector aiming parallel microarray spot processing

t_n			t_{n+1}			t_n
T	K	Input i	$C_2^{(0)}$	$C_1^{(0)}$	$C_0^{(0)}$	Output v
1	1	a_6	a_6	0	0	a_6
2		a_5	a_5	a_6	0	a_5
3		a_4	$a_4 + a_6$	a_5	a_6	a_4
4		a_3	$a_3 + a_5 + a_6$	$a_4 + a_6$	a_5	a_3
5	2		0	$a_3 + a_5 + a_6$	$a_4 + a_6$	$a_2 = a_4 + a_5 + a_6$
6			0	0	$a_3 + a_5 + a_6$	$a_2 = a_3 + a_4 + a_5$
7			0	0	0	$a_2 = a_3 + a_5 + a_6$

simulation time t_n is given by the register cells C_1 + C_0 from the simulation time t_n. This leads to the coding equations described by Eq. (2.35).

$$\begin{cases} a_2 = a_4 + a_5 + a_6 \\ a_1 = a_3 + a_4 + a_5 \\ a_0 = a_3 + a_5 + a_6 \end{cases} \tag{2.35}$$

The VHDL code for the cyclic encoder is described next. The input output ports are detailed in the entity description as presented next.

```
———————— Entity declaration————————————————————————
        entity RD1 is
        Port (  clk:     in std_logic;
                inf:     in std_logic;
                reset:   in std_logic;
                k_switch: in std_logic;
                data:    out  std_logic) ;
        end RD1;
———————— End of entity declaration—————————————————
```

The information bits *i* are sequentially delivered to the *inf* input port, which is connected internally to the first register cell C_2. The *clk* and *reset* inputs represent the clock signal input port and the reset port, respectively. The reset port initializes the register cell C_2, C_1 and C_0 with 0 logic. The *data* output ports delivers sequentially each bit of the v codeword. Note that the size of v is 7 in our example (i.e. C(7, 4) cyclic code). The input port *k_switch,* as its name reveals, represents the switch K. This input port is 1 logic for the first 4 clock cycles, meaning the switch K is on position 1 and the information bits are loaded in the register cells. The next 3 clock cycles the *k_switch* is 0 logic, meaning the input into the shift register is 0 and the control symbols are computed and delivered at the *data* output port.

```
———————— Behavioral description of the cyclic encoder ————————
        architecture Behavioral of RD1 is
        signal temp:STD_LOGIC_VECTOR(2 downto 0);
        signal outt:std_logic;
        signal outt2:std_logic;
        begin
        process (clk)
        variable RDin : std_logic;
        begin
                if k_switch = '1' then
                        RDin : = inf;
                else
                        RDin : = '0';
                end if;
```

```
21:   if reset = '1' then temp <= "000";
22:   elsif clk'event and clk = '1' then
23:   temp <= RDin & temp(2 downto 1);
24:   end if;
25:   end process;
26:   outt <=inf;
27:   outt2 <=temp(0) xor temp(1);
28:   data <= outt when k_switch = '1' else
29:      outt2 ;
30:   end Behavioral;
```

——— The end of the behavioral description of the cyclic encoder ———

The functionality of the decoder is described by the aforementioned behavioral description. There is a *temp* signal declared, which represent the shift register. The register is instantiated through the sequential process description where register cell values are instantiated on clock event (code line 22 and 23). On *reset* (code line 21), the *temp* register in initialized with 0 logic values. The variable *RDin* represents the input into the first register cell, which is a multiplexed input. In case the *k_swith* is 1, the input into the register is the information bits, whereas the *k_switch* is 0, the input is 0. This multiplexed input is described by the code lines 16 to 20. Outside the process there are 3 other concurrent statements, code lines 26, 27 and 28. These code lines connect the output port either to the input through the *outt* wires in case the *k_switch* is 1, or to the *temp(0)* xor *temp(1)* in case the control bits are computed and the k_switch is 0 logic. Note that *k_switch* 1 logic means the switch K is on position 1 and *k_switch* 0 logic means that the switch K is on position 2, as referred to Fig. 2.9.

Figure 2.10 details the functionality of the cyclic encoder through two examples. Let us consider two information sequences given by the following information bits: i_1 = *[1010]* and i_2 = *[0001]* for each of the previously mentioned examples. The information bits i_1 and i_2 are sequentially delivered to the encoder starting with

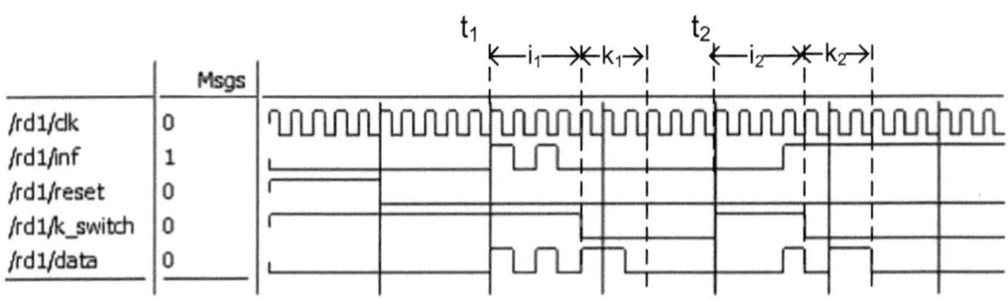

Fig. 2.10 Simulation results for the cyclic encoder for the C(7,4) cyclc code with g (x) = $x^3 + x + 1$

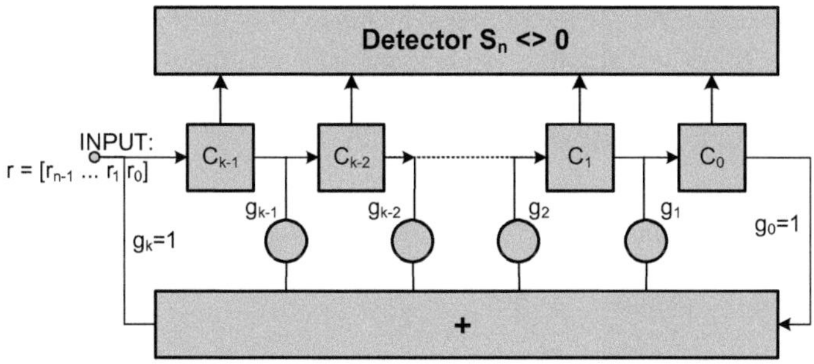

Fig. 2.11 Logic block for the cyclic decoder

simulation time t_1 and t_2 respectively, according to Fig. 2.10. After 4 clock cycles, the number of $k = 3$ control bits are delivered to the output for the next 3 clock cycles. Consequently, the codewords delivered by the encoder at the output corresponding to the two information bits sequences are v1 = [1010110] and v2 = [0001011].

2.6.1 Cyclic Decoder Architectures

Let us consider r to be de received codeword at the decoder side. For error detection, the non-zero state of the shift register shows that an error occurred during information transmission. Consequently, the architecture for the decoder is build based on the encoder, by adding a detector block which specifies if the shift register state S_n is non-zero. This leads to the logic block for the cyclic decoder illustrated in Fig. 2.11. During n clock cycles, the decoder detects based on the register state $S_n = [C_{k-1} \ldots C_0]$ if there are errors during transmission.

In case error correction is desired, based on the S_n syndrome value, the positions of the errors within the received codeword r are computed using supplementary n clock cycles. Thus, during 2n clock cycles the error can be corrected. Detailed architectures for error correction cyclic decoders are presented in [2].

2.7 Conclusions

The current chapter illustrates how channel coding is used for error protection through error detection or correction during data transmission. Basic notions for channel coding such as error control codes representation and parameters are presented. The chapter continues with the description of linear block codes, namely hamming and cyclic codes. Once all the necessary information on error control codes are provided, coder and decoder architectures are presented together with

VHDL codes sequences for their implementation. The functionality of the proposed architectures is shown through simulation. The examples of VHDL code for encoder and decoder implementation represent the starting point for any other implementation of coder/decoder architectures.

References

1. R.G. Gallager, Principles of Digital Communications I, Course materials 2006. MIT OpenCourseWare (http://ocw.mit.edu/), Massachusetts Institute of Technology
2. M. Borda, *Fundamentals in Information Theory and Coding* (Springer, Berlin, 2011)
3. C.E. Shannon, A mathematical theory of communication. Bell Syst. Techn. J. **27**, 379–423 (1948)
4. S. Lin, D. Costello, Error control coding, Prentice-Hall, (1983)
5. G. Cullmann, *Codage et transmission de l'information* (Editions Eyrolles, Paris, 1972)
6. J. Liesen, M. Volker, *Linear Algebra* (Springer, Berlin, 2015)

Chapter 3
High-Throughput Hardware Architecture for LDPC Decoders

The error probability considering communication channel such as electromagnetic wave through space is high. This is why channel encoding for error correction is commonly used. Powerful error correction codes are available (i.e. low-density parity check codes LDPC), unfortunately, the more powerful the codes are, the more computational cost for the decoding at destination increases. Nevertheless, high performance computing architectures (e.g. graphic processing units, FPGAs) are available for implementing high-throughput decoders. This chapter focuses on the implementation of high-throughput LDPC decoders using field programmable gate array (FPGA) technology.

The present chapter starts with a short introduction LDPC codes in digital communication and continues with the description of a low complexity belief propagation based decoding algorithm for LDPC codes. In spite of the iterative nature of the decoding process, the proposed decoding algorithm provides both reduced complexity and increased BER performance as compared with the classic min-sum algorithm generally used for hardware implementations. Linear approximations of check-nodes update function are used in order to reduce the complexity of the belief propagation algorithm. Considering the proposed low-complexity decoding approach, FPGA based hardware architecture is proposed for implementing the decoding algorithm, aiming to increase the decoder throughput. FPGA technology was chosen for the LDPC decoder implementation, considering its parallel computation capabilities and its ease of reconfiguration. The obtained results regarding computational complexity, decoding throughput and BER performances are presented.

B. Belean, *Application-Specific Hardware Architecture Design with VHDL*, Signals and Communication Technology,
https://doi.org/10.1007/978-3-319-65025-8_3

3.1 Introduction to LDPC Codes for Digital Communication

Reduced error probability and high-throughput at destination are mandatory for high speed networks of nowadays digital communications. The aforementioned premises are necessary for any efficient information transmission system considering its accuracy, throughput and computational cost. Thus, the purpose of the channel coding blocks is two folded. On one hand, they assure reliable transmission by performing error detection and correction and on the other hand, the channel encoder and decoder have the function of matching the source to the transmission channel.

For digital communication systems, the accuracy degree is estimated using the bit-error-rate (BER) measure, which corresponds to the probability of an erroneous bit within the received codeword r. One can say BER reflects the transmission quality. A simple interpretation of the BER is the number of erroneous bits divided by the total number of bits transferred during the time interval under analysis. BER is commonly expressed as a percentage representing the likelihood of an erroneous bit.

Low density parity check (LDPC) codes are a class of linear block codes for er-ror correction in communication channels. Introduced by Galager in 1962 [1], LDPC codes offer remarkable performances falling 0.04 dB short of the Shannon theoretical limit, whereas the complexity of their decoding process grows only linearly with block length.

Despite their advantages, LDPC codes were forgotten for the next decades, due to insufficient computational power available for the decoding process. In the last decade, making use of the increased computational power offered by graphic processing units, FPGA/ASIC technologies and digital signal processors, LDPC codes are considered a significant breakthrough in the world of digital communications. Thus, communication standards like WiMAX for wireless networks and DVB-S2 for satellite broadcasting services use LDPC codes, taking into account their remarkable bit error rate (BER) performances. The disadvantage of using LDPC codes consists in the increased computational time for the iterative decoding process, due to the high order block lengths (50 kb for the DVB-S2 standard). An open subject in current research is to improve the LDPC decoder throughput [2, 3]. The state of the art solutions used to overcome the last mentioned disadvantage involve decoding algorithms parallelization using application specific hardware architectures. Existing approaches used to increase the decoder throughput are summarized as follows:

- GPU (Graphic Processing Units) [4–6] exploits data parallelism using multi-thread capabilities and parallel memory access producing performances for LDPC implementations comparable with ASIC dedicated decoders [5];

- ASIC/FPGA technology is used to develop application specific hardware architectures and processing units in order to parallelize LDPC decoding process [7, 8];
- The LDPC decoding process is parallelized using Application Specific Instruction Processors [3].

Further on, the main concepts of LDPC codes together with existing approaches for implementing the LDPC decoder are presented. An *N* dimension LDPC code represents a linear block code whose codeword of length *N* satisfy a set of *M* linear parity-check constraints and its code rate is $R = 1\text{-}M/N$. LDPC codes are defined by the sparse parity-check matrix *H(M,N)* composed by *1* and *0* values, each line specifying one of the parity-check constraints. Thus, each codeword *c* satisfies the relation (3.1) where H^t is H transposed.

$$c * H^t = 0 \tag{3.1}$$

Using the parity-check matrix $H = [Pt\ I]$, the generator matrix *G* for LDPC codeword is obtained ($G = [I\ P]$). The LDPC decoding is characterized by passing probabilistic messages between two types of nodes, *M* check-nodes (CN) and *N* variable nodes (VN), according to Tanner graph [9], (Fig. 2.1). The messages transmitted iteratively between the two types of nodes contain both the probabilities for the node *i* to be *0* and *1*, expressed as a Log-Likelihood Ratio denoted by Eq. (3.2), where $P(y_i = 0)$ denotes the probability of the variable node *i* to be *0*.

$$LLR_i = \log(P(y_i = 0)/P(y_i = 1)) \tag{3.2}$$

Each type of nodes collects incoming messages from its complementary nodes and produces an outgoing message (Fig. 3.1b, c). The message $q_{i,j}$ is sent from VN to CN and represents the probability for the y_i bit from the codeword to be *0* or *1*. The CN sends the $r_{i,j}$ message back to the VN, indicating the probability for the y_i codeword bit to be *0* or *1*, considering the parity check constraints. Thus, the LDPC decoding determines the correct codeword based on the resulted y_i bits, after performing iterative updates of the VN and CN nodes.

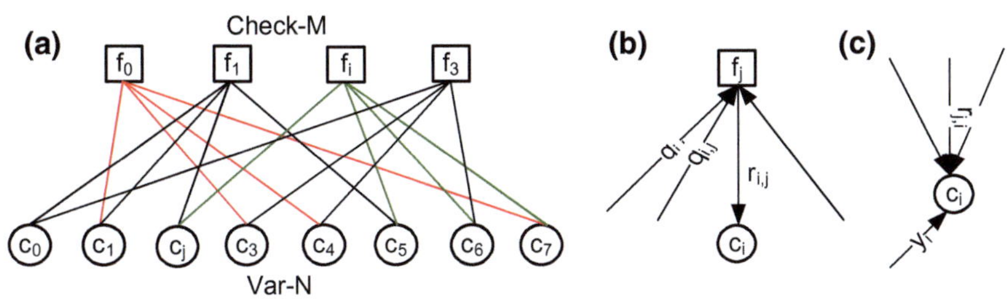

Fig. 3.1 Tanner graph

3.2 Decoding Algorithms Description

The open problem in recent research regarding the iterative LDPC decoding algorithms is to reduce their computation complexity while maintaining the same decoding efficiency (BER) [10]. The most widely used decoding algorithms are the belief propagation and min-sum algorithms denoted BP and, respectively, MS.

Belief propagation algorithm

The BP algorithm [11, 12], summarized in Fig. 3.2, starts with the *M* check nodes and the *N* variables nodes initialization. The received codeword sent through the AWGN channel is $y_n = s_n + w_n$, with s_n the corresponding transmitted sequence and w_n independent Gauss random variables. Thus, the variable nodes are initialized with the a priori log-likelihood ratios of the received codeword y_n. Check nodes (CN) and Variable nodes (VN) updates are performed as described in Fig. 3.2, taking into account the parity check matrix *H* and the way of passing the messages from CN to VN described in Sect. 3.1. The *i* and *j* indices are specified by the position of *1* values from the H matrix rows and, respectively, columns. The decoded codeword is found after performing all iterations in the variable nodes VN.

Min-sum algorithm

In practice, aiming efficient hardware implementations, a low complexity decoding algorithm is used for LDPC codes, known as the min-sum (MS). The aforementioned algorithm replaces the check node update described by Eq. (3.1) by a minimum function [13, 10]. We denote by *x*, the most likely vector representing a code-word such that $H \cdot x = 0$. The received codeword, sent through an AWGN channel, to be decoded is y_n, while r_n represents the a priori log-likelihood ratios of y_n. The *m* terms check sum evaluated from the hard decisions x_m is denoted by σ_m, while/σ_m represents its modulo-2 complement. The decoding algorithm is presented in Fig. 3.3. *Step 1* performs the check-nodes updates; *step 2* performs the variable nodes update, whereas *step 3* represents a stop condition in case the correct codeword is determined before all iterations end. The BER performance of the aforementioned decoding algorithms can be depicted in Fig. 3.4.

Fig. 3.2 Belief propagation algorithm

Belief propagation decoding algorithm

Nodes initialisation
  Check nodes CN <= 0
  Variable nodes VN <=LLR (y_n)
Iterate for k=1,2,...k_{max}
  for each CN(m = 1 to M):
  update $CN = -2\tanh^{-1}(\prod_i \tanh \frac{-VN_{prev}(i) + CN_{prev}(i)}{2})$ (1)
  for each VN(n = 1 to N):
  update $VN = LLR(y_n) + \sum_j CN(j)$ (2)

Fig. 3.3 Min-sum algorithm

Min-sum decoding algorithm
Nodes initialisation x_n <= hard decision of y_n VN: r_n <= LLR (y_n) Iterate for k=1,2,...k_{max} Step1 for each CN(m): $\sigma_m = (\sum x_n)$ ***mod** 2* $y_{mn\ min} = \min \{\lvert y_{n'}\rvert\}$ Step2 for each VN(n): $z_n = \lvert r_n\rvert + \sum(\overline{\sigma_m} - \sigma_m)\, y_{mn\ min}$ Step3 if $z_n < 0$ then $x_n = x_n$ ***xor** 1* $y_n <= z_n$

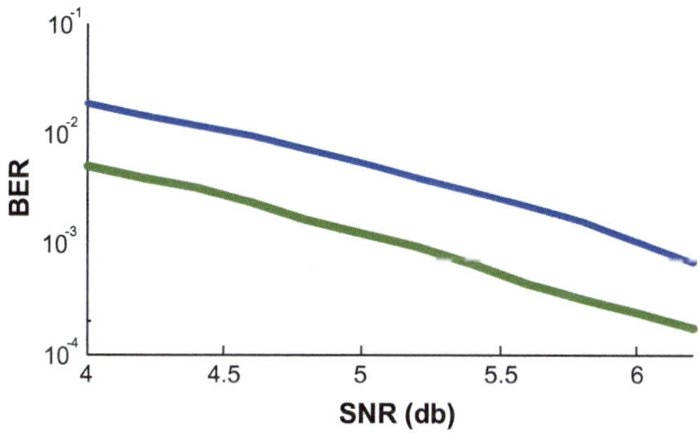

Fig. 3.4 BER performances of BP and MS algorithms

The MS algorithm brings up low computational complexity for the CN update, achieved using a minimum function [2, 14]. Its main disadvantage is the increased bit error rate compared with the BP algorithm used in AWGN communication channels.

3.3 Low-Complexity Approach for LDPC Decoding Process

The importance of implementing low-complexity LDPC decoding algorithms resides in the need of feasible decoder implementations aiming to fulfill the need of high-throughput and increased bit error rate in digital communication, where increased amount of information per time unit is requested yearly. Efficient soft decision decoding algorithms based on density evolution of the belief propagation are proposed in [8, 15, 16]. Moreover, the construction of the sparse parity check matrix H can be taken into account [17] in order to achieve efficient decoding.

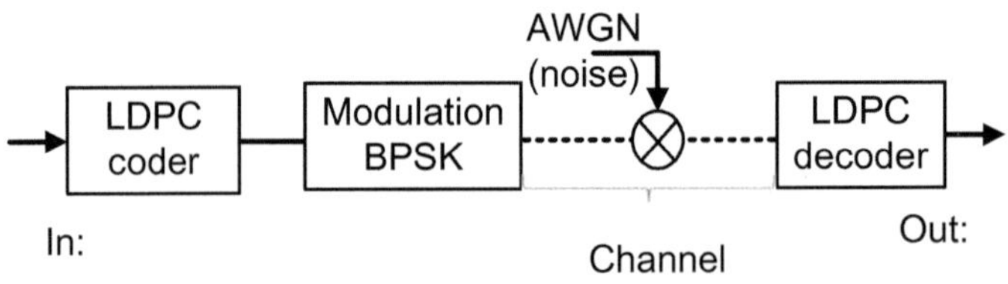

Fig. 3.5 Simulated communication channel

Starting from the same belief propagation decoding scheme, in the current section, a low complexity decoding algorithm LA-BP is presented together with its proposed high-throughput hardware architecture. Experimental results for both decoding algorithm and its hardware implementation are compared with the state of the art approaches and presented in the end of Sect. 3.3. The LDPC decoding algorithms are simulated within an AWGN communication channel with BPSK modulation as shown in Fig. 3.5.

LA-BP decoding algorithm

The proposed decoding algorithm LA-BP is based on the original BP algorithm [11, 12], taking into account its increased BER performance. The complexity of the aforementioned algorithm (section I.B) is given by the iterative update of the check nodes described by the product from Eq. (3.1), which involves increased computational time and increased hardware resource usage. Therefore, as presented in [10], in order to reduce the computational complexity, Eq. (3.1) is written as the sum denoted by (3.3).

$$CN = \prod_i sign(-\lambda(i))f\left(\sum_i f(|\lambda(i)|)\right) \tag{3.3}$$

where

$$f(x) = \log\frac{(e^x + 1)}{(e^x - 1)} \tag{3.4}$$

and

$$\begin{cases} \lambda(i) = CN_{prev}(i) - VN_{prev}(j), \\ \{j \in \overline{1..N} \mid H(j,i) = 1\} \end{cases} \tag{3.5}$$

The $\prod_i sign(-\lambda(i))$ represents the sign for each check node CN update, while updating the check nodes values involve f function computation. Aiming an efficient FPGA based hardware implementation of the LDPC decoder, the computation of f is performed using linear approximations.

FPGA based decoder implementation

FPGA technology uses pre-built logic blocks and programmable routing resources for configuration and for implementing custom hardware functionality. Its main benefits are the low cost, the short time to market and the ease of reconfiguration. Moreover, FPGA technology exploits spatial and temporal parallelism aiming algorithm parallelization for fast processing. These advantages can be used to implement high throughput LDPC decoder architectures [18, 19]. High throughput FPGA based implementation for LDPC decoders are used in various applications, other than wireless communication standards. Thus, a coding scheme for the Gauss wiretap channel based on LDPC codes is proposed in [20]. FPGA based implementation for reduced code length LDPC decoders can also be used in order to correct errors occurred in data storage using flash memories [21]. Efficient hardware architectures for the LDPC decoders bring up both increased throughput and quality of audio and video wireless data transmission in applications as the ones presented in [22] and, respectively, [23]. LDPC codes found also their application in patterned media storage as shown in [24], where the major contribution is a stop update criterion for the LDPC decoding algorithm which leads to an increased LDPC decoder throughput.

The first logic block within the LDPC decoder is the *f* function computation unit (FCU). The FPGA hardware implementation for the FCU is based on linear approximation of the *f* function Eq. (4) and is efficiently designed according to algorithmic constraints (fixed point requirements). Thus, a 16 bit (Q1.12) fixed point representation is used for both the input and output data. The input x represents the log-likelihood ratios of the received symbols y_i, denoted by $LLR(y_i) = 2\cdot y_i/\sigma^2$, where $\sigma = 1/(2R\cdot 10^{EBN0/10})$. The *f* function is calculated in an $i = 1 \ldots n$ points $A_i(x_i, y_i)$, together with the slope m_i for each line described by two adjacent points. Thus, the *f* function is represented by n segments, with $n = 100$ empirically determined to suit the decoder hardware implementation.

Similar approach which uses approximation over consequent intervals were used in [25] for check nodes update. The parallel computation capabilities of the state of the art technologies offer the possibility to increase the number of interval used for approximation. Our proposed hardware implementation for *f* function computation makes use of a LUT based ROM in order to store the A_i and m_i values. Thus, as presented in the next paragraphs, only with the cost of approximately 1 kb of ROM memory, a significantly increased number of segments can be used for approximation. Figure 3.6 illustrates the efficiency of both the approach described in [25] and the proposed one, for approximating the $g(x) = log(1 + e^{-|x|})$. The maximum deviation of our proposed approach for $g(x)$ estimation is also represented by the line marked with squares.

The FCU hardware architecture for $f(x)$ computation is presented in Fig. 3.7. The *f* function values $A_i(x_i, y_i)$ are stored in the "*ROM f values*" memory. Also the slope m_i for each line described by two adjacent points A_i and A_{i+1} is stored in the "*ROM Slope*" memory. Thus, the *f* function is represented by n segments, each of them

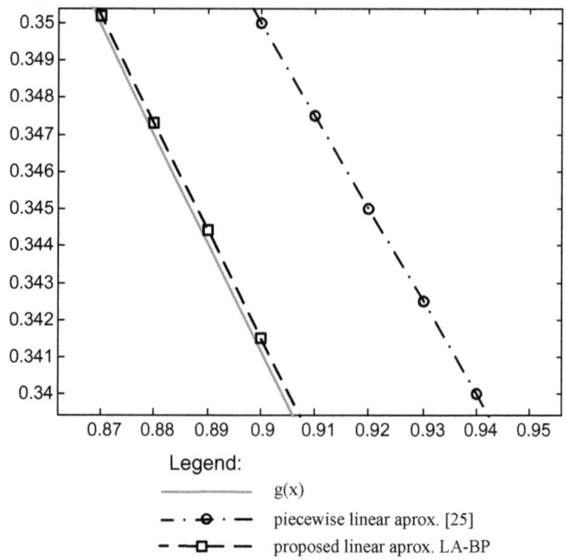

Fig. 3.6 Efficiency estimation of the proposed linear approximation approach compared to the one presented in [25]

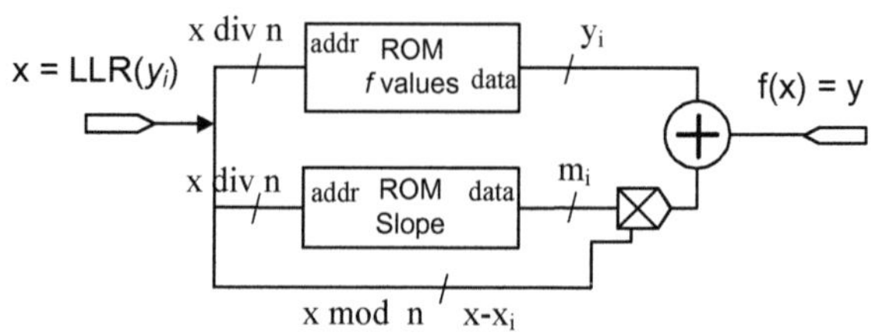

Fig. 3.7 FCU unit

described by the equation of a line with slope m_i which passes through the point $A_i(x_i, y_i)$. In order to compute the f function for a given value x, we are using Eq. (3.6).

$$f(x) = m_i(x - x_i) + y_i \tag{3.6}$$

The offset y_i and the slope m_i are given by the values from the "*ROM f values*" and "*ROM Slope*" respectively at *x div n* address. The *x-x_i* values are given by *x mod n*. It is to be mentioned that for an efficient implementation, *n* value is chosen as a power of *2*. The next code example illustrates how linear approximations are used to compute *f(x)* values.

Example 3.1—Hardware architecture for f(x) computation used in check-nodes updates

This example describes the computation unit for *f(x)* denoted by Eq. (3.6) using two the ROM memories, namely *ROM_slope* and *ROM_fvalues*. The input *x* represents the log-likelihood ratios of the received symbols y_i, denoted by LLR(yi)

Table 3.1 ROM_slope and ROM_fvalues memory values for f(x) approxmation

	ROM Slope		ROM fvalues		
ROM memory address i	**Slope m_i** (real value)	**Slope m_i** (fix point representation)	**Offset y_i** (real value)	**Offset y_i** (fix point representation)	x_i
1	4.809702	X"4CF4"	1.505291	X"1815"	0.0625
2	2.806184	X"2CE6"	1.204685	X"1346"	0.125
3	1.983276	X"1FBB"	1.029298	X"1078"	0.1875
4	1.530391	X"187C"	0.905344	X"0E7C"	0.25
128	0.000301	X"0001"	0.000291	X"0001"	0.00029

$LLR(y_i) = 2yi/\sigma^2$, where $\sigma = 2R10^{E_B N_0/10}$. Considering a code rate of $R = 0.35$ and the signal/noise ratio levels $E_B N_O$ = *[4db...8db]*, *f* is defined on the range *(0... 8)* of real numbers.

A fixed point data representation *Q(4,12)* was chosen for the *f* function implementation used for check nodes updates. Thus, fixed point values are delivered as *x* = *Y(15:0)* inputs which have a 4 bits integer part and 12 bits fractional part. Similarly, the computed *f(x)* values are represented using the same precision of 12 bits for the fractional part. For each values *x*, delivered at the input, we determine the ROM memory location (address) by computing the integer part of the division *x/n (i.e. x* div *n)*, where *n* represents the number of segments used for linear approximation of function *f(x)*. The content of the *ROM_slope* and *ROM_fvalues* at the computed address (*x* div *n*) contain the slope and offset denoted by m_i and y_i, respectively, used by Eq. (3.6) to compute the *f* function value for its argument *x*. Considering the Eq. (3.6), the value denoted by *(x-x$_i$)* remains to be computed in order to deliver the *f(x)* value. This *(x-x$_i$)* value is given by the reminder of the x/n division, (i.e. *x* mod *n*). The content of the *ROM_slope* and *ROM_fvalues* used for linear approximation of *f(x)*, considering a number of *n* = *128* linear segments is illustrated in Table 3.1.

The VHDL code for the *f(x)* computation unit depicted in Fig. 3.8 is presented next. As detailed in Fig. 3.7, the logic unit is based on two ROM memories. First the Vhdl code for ROM memory declaration is presented, followed by the VHDL description of the *f(x)* unit, which includes as components the two ROM memories containing the slopes and offsets of the segments used for linear approximations. Thus, the entity declaration for the ROM memories is as follows:

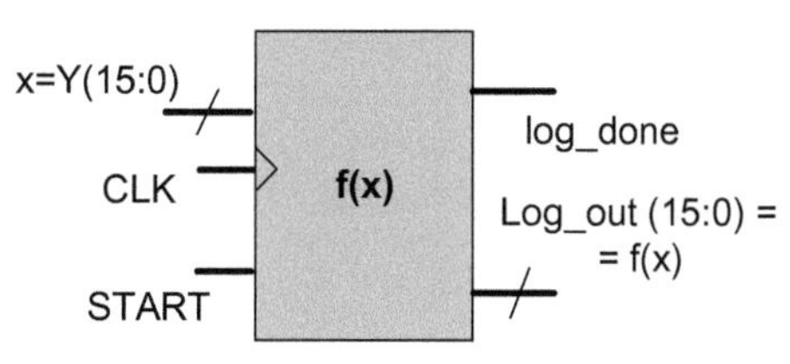

Fig. 3.8 Logic block for the ROM based f(x) computation unit

```
------------------------- Entity declaration-------------------------
 entity ROM_LOG is
         Port ( ADDR : in  STD_LOGIC_VECTOR (6 downto 0);
                    LOG_POINT : out  STD_LOGIC_VECTOR (15 downto 0);
                    LOG_SLOPE : out  STD_LOGIC_VECTOR (15 downto 0));
end ROM_LOG;
------------------------- End of entity declaration-------------------------
```

Once the inputs outputs of the ROM memroes are defined using the entity, we can specify the behavior of the logic unit. In this case, the content of ROM memories is specified by two constants *ROM_LOG* and *ROM_SLOPE,* and ROM memory content is addressed through the address *ADDR*, delivered at the input.

```
------------------------- Behavioral description of ROM Memories-------------------------
architecture Behavioral of ROM_LOG is

signal ADDR_INT:  integer range 0 to 127: = 0;
type ROM_DATE_LOG is array (0 to 127) of std_logic_vector (15 downto 0);
type ROM_DATE_SLOPE is array (0 to 127) of std_logic_vector(15 downto0);
constant ROM_LOG: ROM_DATE_LOG: =
        ("0001100000010101",      "0001001101000110", "0001000001111000",...
constant ROM_SLOPE: ROM_DATE_SLOPE:=
("0100110011110100",       "0010110011100110", "0001111110111011",...

LOG_POINT  <= ROM_LOG (conv_integer(ADDR));
LOG_SLOPE  <= ROM_SLOPE(conv_integer(ADDR));
end Behavioral;
```

It can be noticed there is no clock signal, which means the ROM memories architectures are fully combinational, as expected.

The *f(x)* computation is performed trough the logic unit *ROM_LOG* described next. With regards to the "black-box" principle for building logic units, the inputs and outputs of the *ROM_LOG* are: *clk, START, Y* representing the inputs and *Done*, *Log_out* representing the outputs. *START* signalizes the beginning of the computation, whereas *Done* marks when the result for the input *Y* is available at the output.

```
------------------------- Entity declaration-------------------------
entity Log_Core is
Port  (CLK : in   STD_LOGIC;
       START: in STD_LOGIC;
       Y : in  STD_LOGIC_VECTOR (15 downto 0);
       Log_out : out  STD_LOGIC_VECTOR (15 downto 0);
       Done : out  STD_LOGIC);
end Log_Core;
------------------------- End of entity declaration-------------------------
```

This logic block uses as components the two ROM memories for *f(x)* computation. The component declaration is done before the architecture *begin*, using the following syntax: *component ROM_LOG*. The *ROM_LOG* component instantiation is performed by the following code line: *ROM_LOG_INST: ROM_LOG*. The *PIPE_LEVEL* constant specifies the number of clock periods used by the architecture to deliver the result corresponding to its input *Y*. The signals *Y_LOW_DEL, Y_MULT, LOG_RES, ROM_ADDRESS, LOG_POINT, LOG_SLOPE, Ysig* are used as wires for the interconnections betweens multipliers, adders and ROM memories used within the *f(x)* computation unit.

The processes *sum_for_result*, *generate_done* and *del_addr* describe how the computation unit works. The *del_addr* process computes the x- $x_i = x \bmod n$ value used to address the ROM memories. The computed address specifies the ROM memories contents given by *LOG_POINT* and *LOG_SLOPE* used to compute the output value *log_out* for the input *Y*. The *generate_done* process marks when the result is available at the output *log_out*. The actual *f(x)* computation is performed by the concurrent statement *Y_MULT <= Y_LOW_DEL * LOG_SLOPE* and the process *sum_for_result*. Consequently, the final result is given by the *LOG_RES* signal wired to the output *log_out*, and valid once the *done_pipe* signal marks the end of the computation.

```
------------------- Architecture description------------------------
architecture Behavioral of Log_Core is
CONSTANT PIPE_LEVEL : integer: = 2;
signal Ysig: std_logic_vector (15 downto 0) : = X"0000";
signal Y_LOW_DEL: std_logic_vector (8 downto 0): = '0'&X"00";
signal Y_MULT: std_logic_vector (18 downto 0) : = "000"&X"0000";
signal LOG_RES: std_logic_vector (19 downto 0) : = X"00000";
signal ROM_ADDRESS: std_logic_vector (6 downto 0): = "0000000";
signal LOG_POINT : STD_LOGIC_VECTOR (19 downto 0): = X"00000";
signal LOG_SLOPE : STD_LOGIC_VECTOR (9 downto 0): = "00"&X"00";
signal done_pipe: std_logic_vector (Pipe_level-1 downto 0): = (others =>0);
component ROM_LOG
      Port ( ADDR : in STD_LOGIC_VECTOR (6 downto 0);
        LOG_POINT : out STD_LOGIC_VECTOR (19 downto 0);
        LOG_SLOPE : out STD_LOGIC_VECTOR (9 downto 0));
end component;
begin
del_addr: process (CLK, Y)
begin
    if CLK'EVENT and CLK = '1' then
      ROM_ADDRESS <= Y(15 downto 9);
      Y_LOW_DEL <= Y(8 downto 0);
    end if;
end process del_addr;
```

```
ROM_LOG_INST: ROM_LOG
        Port map ( ADDR => ROM_ADDRESS,
                   LOG_POINT => LOG_POINT,
                   LOG_SLOPE => LOG_SLOPE);

Y_MULT <= Y_LOW_DEL * LOG_SLOPE;

sum_for_result:process (CLK, Y_MULT, LOG_POINT)
begin
     if CLK'EVENT and CLK = '1' then
      LOG_RES <= Y_MULT + LOG_POINT;
     end if;
end process sum_for_result;
generate_done: process (CLK, START, done_pipe)
begin
     IF CLK'EVENT and CLK = '1' then
        done_pipe (PIPE_LEVEL - 1 downto 0) <= done_pipe (PIPE_LEVEL – 2
                                               downto 0) & START;
     end if;
end process generate_done;

Log_out <= LOG_RES;
Done <= done_pipe (PIPE_LEVEL-1);
end Behavioral;
------------------------ End of the architecture description--------------------
```

The previous VHDL example describes the functionality of the logic block used for *f(x)* computation using linear approximations. Further on, the second example shows how the designed logic unit is integrated using a structural description within a vhdl test-bench, in order to test its functionality. The test-bench delivers in a sequential manner input data to the designed logic block and displays the corresponding output signals.

Example 3.2—Test-bench for the f(x) computation unit

The unit (i.e. test-bench) used to test the functionality of the designed logic block for *f(x)* computation is called *test_log_core*. As it can be seen further on, there are no inputs or outputs defined within the entity corresponding to the proposed test-bench.

```
---------------------- Entity declaration---------------------------------------
ENTITY Test_Log_Core IS
END Test_Log_Core;
-------------------- End of entity declaration-----------------------------------
```

Within the architecture test-bench, the component representing the logic unit for f(x) computation is defined, as presented next.

```
———————————————— Architecture description————————————————
ARCHITECTURE behavior OF Test_Log_Core IS
------ Component Declaration for the Unit Under Test (UUT)
      COMPONENT Log_Core
      PORT(
             CLK : IN std_logic;
             START: IN std_logic;
             Y : IN std_logic_vector(15 downto 0);
             Log_out : OUT std_logic_vector(19 downto 0);
             Done : OUT std_logic      );
      END COMPONENT;
```

The inputs and outputs are defined next as architecture defined signals, which will be used to deliver inputs and to collect the outputs signals of the designed logic block for *f(x)* computation. Consequently, the signals are connected to the logic block under test when the component associated to the logic bloc is instantiated. See the following section of VHDL code.

```
———————————— Architecture description continued————————————
--Inputs
   signal CLK : std_logic : = '0';
   signal Y : std_logic_vector(15 downto 0) : = (others => '0');
   signal START : std_logic: = '0';
---Outputs
   signal Log_out : std_logic_vector(19 downto 0);
   signal Done : std_logic;
   constant CLK_period: time: = 10 ns;
   signal log_res: real;

BEGIN
uut: Log_Core PORT MAP (
            CLK => CLK,
            START => START,
            Y => Y,
            Log_out => Log_out,
            Done => Done);

CLK_process :process
begin
      CLK <= '0';
      wait for CLK_period/2;
      CLK <= '1';
      wait for CLK_period/2;
end process;
```

```
generate_input: process
begin
        wait until CLK'EVENT and CLK = '1';
          Y <= Y  +  X"01" after 2 ns;
end process generate_input;

generate_start: process
      variable i: integer range 0 to 7 : = 0;
      begin
           START <='0' after 2 ns;
          for i in 0 to 5 loop
           wait until CLK'EVENT and CLK = '1';
          end loop;
            START <='1' after 2 ns;
          wait until CLK'EVENT and CLK = '1';
      end process generate_start;
END
———————————— End of the architecture description————————————
```

After the component instantiation, subsequent processes are defined. The *CLK_process* generates the clock signal wired to the input of the *Log_core* component. The *generate_input* process delivers sequential input values *Y* to the logic unit for *f(x)* computation. The *generate_start* process marks the beginning of the *f (x)* computation, whereas the output *Done* is generated by the computation unit when there is available data to the *Log_out* output.

A full pipelined architecture was developed for the proposed implementation to maximize the processing throughput. This choice reduces the computational time for the FCU to 1 pixel/clock cycle with an initial 3 clock cycles delay. The sparse property of the parity matrix and the possibility of parallel update of each variable node and check nodes offer the possibility to design parallel application specific hardware architecture for the LDPC decoders.

Our proposed architecture described in Fig. 3.9a makes use of multiple instances of the FCU unit, for check nodes update. The check nodes and variable nodes values, fixed point representation, are stored in R3 and R4 parallel access registers, having the size *M* and *N*, respectively. The H matrix values are stored in a look-up-table, called H matrix LUT. The update of the check-nodes within one of the decoding iterations is performed as follows. The positions *i* of one value within each *H* line specifies the position of *Variable nodes prev* (R1) register values used to update the current check node *j*. Each accumulator computes both the absolute value $|R2_j - R1(i)|$ and its sign. Multiple instances of the FCU units are used in order to compute in a parallel manner the *f* function of their inputs, with the total cost of T_{S1}, corresponding to the first stage *S1* of the check nodes update. In a similar manner, let T_{S2} and T_{S3} be the computational cost for the second and, respectively the third stage. The total delay path for updating all check nodes within

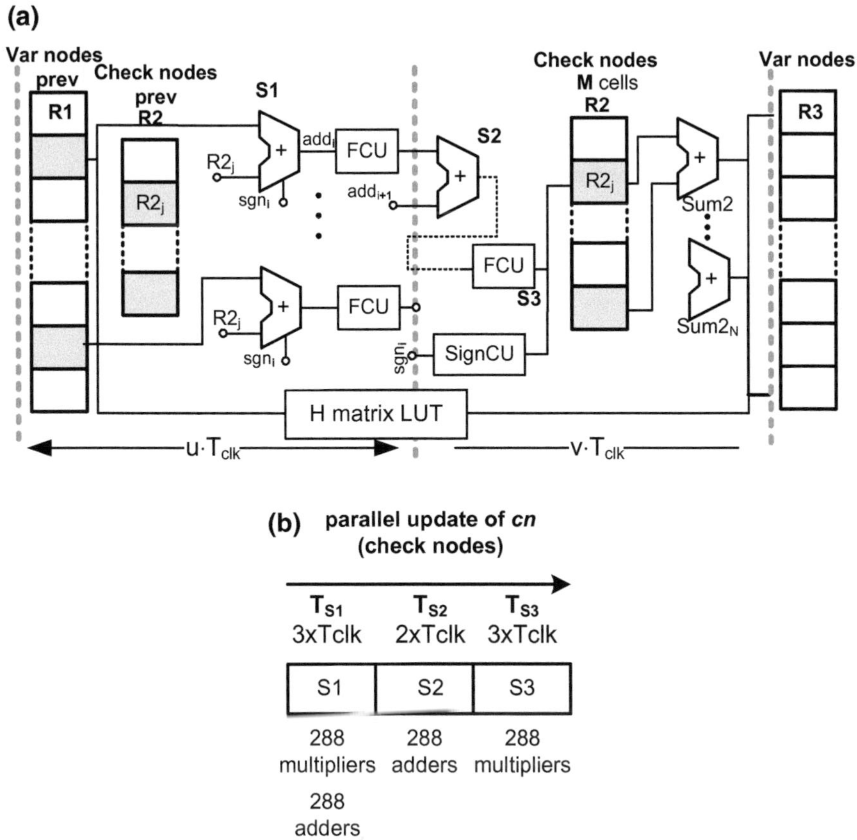

Fig. 3.9 **a** Logic block for the ROM based f(x) computation unit. **b** Logic block for the ROM based f(x) computation unit

one of the decoding iterations is $uT_{clk} = T_{S1} + T_{S2} + T_{S3}$. The variable nodes updates are also performed based on each column values of the *H* matrix and the updated check nodes using vT_{clk} cycles. At the same time, the content of the *R2* register is loaded in parallel with the updated check nodes values (*R3*), and the *R1* content is also updated with the newly computed variable nodes from *R4*. The total delay path for performing one iteration is $T = u+v$ clock cycles. While performing the linear approximation corresponding to the stage S1 and the S2 and S3 stages, the *SignCU* unit computes the sign for each check node update with no additional delay path. Results of the proposed architecture in case of an LDPC decoder implementation are presented in the next section.

The proposed LA-BP approach for implementing the LDPC decoding algorithm, which makes use of linear approximation for check nodes update function, is compared in terms of BER efficiency with the existing min-sum and belief propagation decoding approaches. All the decoding algorithms are implemented using C code and simulated within an AWGN communication channel with BPSK modulation. The LDPC code of size $N = 20$ and $M = 15$ with a code rate of $R = 0.35$ is used for exemplification. The signal/noise ratio levels considered for simulation are $E_B/N_0 = [4db...8db]$ on x axis.

The number of code blocks is 10^6, with the maximum number of performed iterations 30. Fixed point representation $Q(8.16)$ was used for nodes values and 100 approximation segments for the function f. Figure 3.10 illustrates the results obtained by our approach, which are close to the classic belief propagation decoding, and significantly increased compared with the min-sum (MS) algorithm, used in hardware implementation.

In order to assess the decoding throughput in case of the proposed decoder design, an LDPC code of size $N = 576$ and $M = 288$ is used for estimation. The hardware resource usage for the decoder implementation is summarized in the Table I. Let $T = T_{S1} + T_{S2} + T_{S3} + vT_{clk}$, be the total delay path for computing the check nodes and variable nodes within one of the decoding iterations. The data used in the computational stages S1, S2 and S3 are inter-dependent, thus a pipeline strategy cannot be used for parallel computing. Nevertheless, the check-nodes update within one of the iteration are independently computed, thus a group of $cn = 96$ check nodes are updated in a concurrent manner.

In the timing diagram from Fig. 3.9b, the resources for parallel computation of the cn check nodes are listed. The number of iterations considered for the decoding process is $N_{iterations} = 8$. In case of our proposed implementation each line of the parity check matrix contains 6 '1' values, thus the total number of multipliers and adders for $S1$ computing stage is $96 \times 6 = 576$. A fully $k = 8$ stage pipeline architecture is designed for computing the whole $M = 288$ check nodes update. Each pipeline stage introduces a cost of $1\ T_{clk}$ clock cycle and they are described as

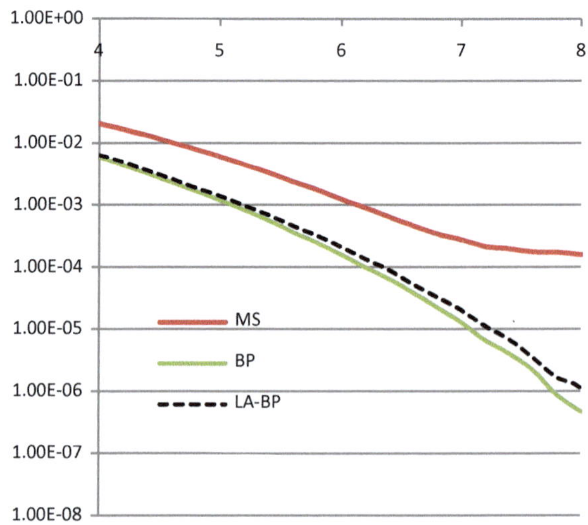

Fig. 3.10 Results of the decoder implementation: BER versus signal/noise ratio

follows: ROM memory data read within S1 stage, the multiplication performed within S1 stage, the addition within S1 stage, two stages corresponding to the S2 additions, ROM memory data read within S3 stage, the multiplication performed within S3 stage, the addition within S3 stage. Thus a total computational time for the check nodes update is $T_{S1} + T_{S2} + T_{S3} = (k + M/cn)T_{clk}$, = = $11\ T_{clk}$, where k is the initial delay due to the pipeline approach. Considering a $2T_{clk}$ computational time needed for the variable nodes update, the number of clock cycles needed by the proposed implementation in case of one of the iterations of the decoding algorithm is computed as follows: $T = T_{S1} + T_{S2} + T_{S3} + vT_{clk}$, leading to a total delay path of *13* clock cycles. For the hardware implementation of the proposed LDPC decoder scheme, the Xilinx FPGA XC71500T was chosen, having a number of 1600 DSP48E. In this way, 1600 multipliers are available for the computation of the decoding algorithm. Table 3.2 presents the hardware resource usage in case of the proposed chip. The aforementioned LDPC code is integrated in the WiMAX standard, thus a throughput comparison with different implementation is performed next.

$$Throughput = \frac{M \cdot f_{clk} \cdot R}{N_{iterations} \cdot T} \quad (3.7)$$

Considering the 400 MHz clock rate, the throughput of the decoder implementation achieves 1,028 Gb/s, which is more than 200 Mb/s higher than the architecture proposed in [2], Table III. Due to the increased BER performances, the number of iterations can be reduced for an increased throughput.

The decoder throughput depends on the level of parallelism used for check-nodes updates, the most time consuming operation within the LDPC decoding process. The most efficient implementations regarding the decoder throughput are obtained using FPGA/ASIC and ASIP approaches. In case of the ASIP approach, [26–28] the level of parallelism depends on the number of microprocessor cores p and registers size n. The microprocessor registers are used to perform parallel updates for pxk nodes, where a node value representation is on n/k bits. In case of FPGA based approaches, the level of parallelism for nodes update is given exclusively by the available FPGA resources. The improvements to the LDPC decoders, compared to existing approaches, introduced by the proposed design are presented in Table 3.3.

Table 3.2 Hardware resource usage of the decoder architecture in case of XILINX FPGA XC71500T

	Used resources		Available resource
Logic units	Number of units	Unit size	Available units
DSP48E	672	24 × 24	1600
Adders	1728	24b	–
LUT based ROM memories	74	1 k x 24b	–
Total slice LUTs	784.400		1.139.200

Table 3.3 Throughput estimation of the state of the art approaches for LDPC decoder implementations

	Code length (standard)	Iterations	f_{clk} (MHz)	Throughput (Mb/s)
Proposed	576 × 288	8	400	1028
[2]	672 × 336	8	335	822
[26]	576 × 288	10	300	240
[28]	576 × 288	10-20	400	237
[25]	IEEE 802.11	8	150	100
[5]	IEEE 802.11	10-50	1300	100

3.4 Conclusions

In a world of digital communication, the LDPC codes bring up a remarkable potential regarding high-throughput channel coding for error correction. Min-sum algorithm is used for efficient hardware implementation for high-throughput LDPC decoder, due to its decreased computation complexity compared with classical belief propagation algorithm. The paper proposes low complexity belief propagation based decoding algorithm which shows increased BER performance compared to the min-sum (MS) algorithm. Moreover, a novel FPGA based hardware architecture is proposed for the low complexity decoding algorithm. Thus, the proposed architecture delivers a throughput up to 1.028 GHz in case of LDPC code size compliant with the WiMAX standard for wireless communication, overcoming the existing implementation in terms of throughput and BER performance. LDPC with reduced codeword are also finding increasing use in applications where reliable and highly efficient information transfer over bandwidth is required. The proposed FPGA based implementation for the LDPC decoder is suitable for this kind of applications due to its parallel computation capabilities, low-cost and ease of reconfiguration.

References

1. R.G. Gallager, Low-density parity-check codes, IRE Transm. Inf. Theory **IT8**, 21–28 (Jan. 1962)
2. S. Kim, G.E. Sobelman, H. Lee, A reduced-complexity archi-tecture for LDPC Layered de-coding schemes, IEEE Trans. VLSI Syst. **19**(6), 1099–1103 (2011)
3. Steffen Kunze, Emil Matus, Gerhard P. Fettweis, *ASIP decoder architecture for convolutional and LDPC codes* (IEEE Int. Symp. Circuits Syst., ISCAS, 2009)
4. H. Ji, J. Cho, W. Sung, Memory access optimized implementation of cyclic and quasi-cyclic LDPC codes on a GPGPU. J. Sig. Process. Syst. **64**(1), 149–159 (2011)
5. G. Falcao, L. Sousa, V. Silva, Massively LDPC Decoding on Multicore Architectures, IEEE Trans. Parallel Distrib. Syst. **22**(2), 309–322 (2011)
6. Houston M, Stanford University, General Purpose Computation on Graphics Processors, http://graphics.stanford.edu/~mhouston/, (2008)

7. J. Su, K. Liu, H. Min, Hardware efficient decoding of LDPC codes using partial-min algorithms, IEEE Trans. Consum. Electron. **52**(4), 1463–1468 (November 2006)
8. Z. Zhang et al., Design of LDPC decoders for improved low error rate perfor-mance: quantization and algorithm choices. IEEE Trans. Wireless Comm. **8**(11), 3258–3268 (2009)
9. F.R. Kschischang, B.J. Frey, H.-A. Loeliger, Factor graphs and the sum-product algorithm. IEEE Trans. Inf. Theory **47**(2), 498–519 (2001)
10. J. Chen, A. Dholakia, E. Eleftheriou, M. Fossorier and X. Hu, Reduced Complexity Decoding of LDPC Codes, IEEE Trans. Commun. **53**, 1232–1232 (2005)
11. Sunghwan Kim, Min-Ho Janget et al., Sequential message-passing decoding of LDPC codes by partitioning check nodes. IEEE Trans. on Communications **56**(7), 1025–1031 (2008)
12. John R. Barry, Low-density parity-check codes, Georgia Institute of Tech., http://www.sce.carleton.ca/~hsaeedi/ldpc.pdf, October 2001
13. X. Zhang, Y. Tian, et al., An multi-rate LDPC decoder based on ASIP for DMB-TH, in *IEEE 8th International Conference on ASIC*, pp. 995–998 (2009)
14. D. Oh, K. Parhi, Min-Sum decoder architectures with reduced word length for LDPC codes, IEEE Trans. Circuits Syst. **57**(1), 105–115 (January 2010)
15. J. Chen, M. Fossorier, Density evolution for two improved BP-based decoding al-gorithms of LDPC codes. IEEE Commun. Lett. **6**(5), 208–2010 (2002)
16. J. Singh, D. Pesch, Application of energy efficient soft-decision error control in wire-less sensor networks, Telecommun. Syst. **52**(4), 2573–2583 (2011)
17. M. Kupimai, A. Meesomboon, V. Imtawil, Low complexity encoder of high rate irregular QC-LDPC codes for partial response channels, Adv. Electr. Comput. Eng. **11**(4), 47–54 (2011)
18. C. Jego, P. Adde, C. Leroux, Full-parallel architecture for turbo decoding of product codes. Electron. Lett. **42**(18), 1052–1054 (2006)
19. R. Terebes, *Mobile communication systems. Part one: GSM networks* (UTPRES, Cluj-Napoca, 2006). ISBN 978-973-662-221
20. D. Kline, J. Ha, S. McLaughlin, J. Barros and B. Kwak, LDPC codes for physical layer security, IEEE Globecom Proceedings (2009)
21. Q. Huang, S. Pan., M. Zhang and Z. Wang, A concatenation scheme of LDPC codes and source codes for flash memories, EURASIP J. Adv. Sig. Process. 2012:**208** (2012)
22. A.D. Potorac, Considerations on VoIP throughput in 802.11 networks, Adv. Electr. Comput. Eng. **9**(3), 45–50 (2009)
23. E. Puschita, P. Kántor, G. Manuliac, T. Palade, J. Bitó, Enabling Frame-Based Adaptive Video Transmission in a Multilink Environment, Adv. Electr. Comput. Eng. **12**(2), 9–14 (2012)
24. X. Liu, J. Cai, L. Wu, Improved decoding algorithm of serial belief propagation with a stop updating criterion for LDPC codes and applications in patterned media storage. IEEE Trans. Magn. **49**(2), 829–836 (2013)
25. X. Hu, E. Eleftheriou, D. Arnold, A. Dholakia, Efficient Implementations of the Sum-Product Algorithm for Decoding, Global Telecommunications Conference Proceedings, Globecom (2001)
26. M. Awais, A. Singh, G. Masera, High throughput LDPC decoder for WiMAX (802.16e) applications, in *Advances in Computing and Communications, Communica-tions in Computer and Information Science*, vol. 191 (2011) pp. 374–385
27. Y. Sun, J. Cavallaro, A flexible LDPC/turbo decoder architecture. J. Sig. Process. Syst. **64**(1), 1–16 (2011)
28. M. Alles, T. Vogt, N. Wehn, FlexiChaP: a reconfigurable ASIP for convolutional, turbo, and LDPC code decoding, 5th International Symposium on Turbo Codes and Related Topics, (2008)

Chapter 4
Hardware Architecture for Edge Detection

Edge detection is a common image processing task which detects the discontinuities in image brightness. Image brightness discontinuities are more likely to correspond to the boundaries of objects within the analyzed image. Consequently, edge detection represents an important image processing tool used in features detection and extraction. In case automation and high-throughput are needed, hardware architectures represent a well-known solution for implementing edge detection algorithms. The present chapter describes an FPGA-based implementation applied in automatic microarray feature extraction. An overview of an automatic microarray image processing and acquisitionsystem is provided, followed by a description of image convolution and its hardware implementation. The architecture for convolution is used to build-up a Canny edge detection filter. The integration of the proposed filter as co-processor architecture within an automatic image processing system is performed using a finite-state-machine (FSM) approach.

4.1 Introduction—Microarray Image Processing System

Measurement of gene expression can provide clues about regulatory mechanism, biochemical pathways and broader cellular function. By gene expression we understand the transformation of gene's information into proteins. The informational pathway in gene expression is: DNA → mRNA → protein. The protein coding information is transmitted by an intermediate molecule called messenger ribonucleic acid mRNA. This molecule passes from nucleus to cytoplasm carrying the information to build up proteins [1]. This mRNA acid is a single stranded molecule from the original DNA and is subject to degradation, so it is transformed into stable complementary DNA for further examination. Microarray technology is based on creating DNA microarrays which represents gene specific probes printed

B. Belean, *Application-Specific Hardware Architecture Design with VHDL*,
Signals and Communication Technology,
https://doi.org/10.1007/978-3-319-65025-8_4

on a glass slide or microchip. The most common use for DNA microarrays is to measure, simultaneously, the level of gene expression for every gene in a genome [2]. In this way, the microarray compares genes from normal cells with abnormal or treated cells, determining and understanding the genes involved in different diseases. The microarray technology is used also in toxicological research and monitoring environmental effects on different genomes.

DNA microarrays which represent gene specific probes arrayed on a matrix such as a glass slide or microchip. Usually samples from two sources are labeled with two different fluorescent markers and hybridized on the same array (glass slide). The hybridization process represents the tendency of 2 single stranded DNA molecules to bind together. After hybridization, the array is scanned using two light sources with different wavelengths (red and green) to determine the amount of labeled sample bound to each spot through hybridization process. The light sources induce fluorescence in the spots which is captured as a gene expression level by a scanner and a composite image is produced [3]. The microarray image represents a collection of microarray spots disposed in a rectangular or hexagonal grid (see Fig. 4.1). Once the expression levels for each microarray spot are estimated, the genes differentially expressed within a microarray experiment are determined and they are named up or down regulated genes. The biologists and medical doctors are interested in the interpretation of the relative changes in intensities for the same spot from the sample and reference image, I_{Cy3} and I_{Cy5} respectively. The selection of differentially expressed genes is done using the fold change value Fc [4], which is given by the log odd ratio off the spot intensity from the two microarray images, sample and reference image. A detailed description of microarray image processing algorithms is presented. The classical flow of processing a microarray image is generally separated in the following tasks: addressing, segmentation, intensity extraction and pre-processing to improve image quality and enhance weakly expressed spots. The first step (i) associates an address to each spot of the image. In the second one (ii), pixels are classified either as foreground, representing the DNA spots, or as background. The last step (iii) calculates the intensities of each spot and also estimates background intensity values. The major tasks of microarray image processing are to identify the microarray image characteristics including the array

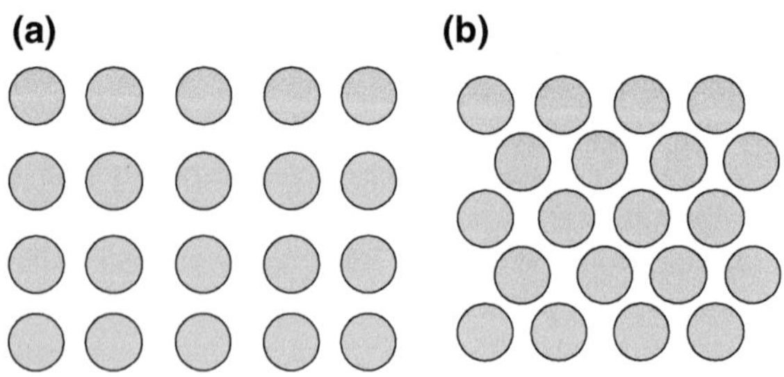

Fig. 4.1 Microarray image classification considering the grid layout: **a** rectangular grid layout, **b** hexagonal grid layout

layout, spot locations, size and shape, and to estimate spot and background intensity values. Existing software platforms (Agilent Feature Extraction Software (FE), GenePix Pro, ScanAlyze) perform the image processing tasks and their results can be found on the public data repository, in case of various microarray experiments (Gene Expression Omnibus—GEO data repository).

The steps of a microarray experiment, from gene specific probes to the biologic interpretation of the results can be summarized as follows: (1) the microscopic plate (microarray) printing, (2) labeling and hybridization of DNA samples, (3) the double laser scanning of the microarray printed plate, (4) the acquisition of the microarray images using a digital camera (5) the image processing algorithms for microarray spots identification, (6) the extraction of microarray spot parameters and biologic interpretation of the results.

The steps (1) and (2) require special laboratory conditions, so they cannot be included in an automatic system for microarray image processing. On the other hand, steps which have as objectives scanning and analysis of microarray images, may be included in an automatic system that can be useful for the medical or military personnel deployed in different locations, or if scientific missions that take place in areas where access to advanced medical equipment does not exist. Steps (3) to (6) describe the methods used in the acquisition and processing of microarray images. They are performed by a bio-information using a microarray scanner together with a workstation. Nevertheless, the increased number of microarray applications made the process of estimating the genes expression levels very important, which leads to the need of an automatic system for r microarray image acquisition and processing. The aforementioned steps can be replaced by a "system-on-a-chip" that eliminates the human intervention and which increases the computation efficiency if algorithms with a high degree of parallelism are employed. FPGA represent a solution for the hardware implementation of image processing specific algorithms which eliminates the shortcomings of the existing software platforms: user intervention, increased computation time and cost. Once automatic microarray image processing algorithms are developed [5], they can be used to build hardware architectures which can be integrated at the scanner level. In this way, gene expression levels are delivered as raw data parameters without the user-intervention. As mentioned before, image processing automation and its integration at the microarray scanner level is motivated by the following scenario: bio-chip technology has multiple applications and its need and use may be compared to the requirement of X-ray radiology in a hospital.

Considering the demand for automation in image processing systems such as microarray analysis, an example of how to develop hardware architecture for a simple image processing task (edge detection) is described further on. Once the hardware architecture is developed for convolution, it is used to build-up a more complex architecture, namely the Canny edge detector, which is in turn integrated as a speed-up co-processor in a micro-processor system. The methodology to integrate the hardware architecture within a microprocessor system is also detailed.

4.2 Hardware Architecture for Image Convolution

4.2.1 Convolution in Digital Image Processing

Digital images represent a two-dimensional (rectangular) array of pixels, each of the pixels having a particular value. Consequently, a digital image can be represented by a function *f(x,y)*, where x and y are the spatial coordinates, and the value of *f* at any pairof coordinates is called the intensity of the image at location *(x,y)*. Computer algorithms are used to perform various operations on digital images; this process is known as digital image processing. In image processing, various tasks such as blurring, sharpening, embossing or edge-detection are accomplished using an operation called convolution. In mathematics, convolution is described as a function that is the integral or summation of two component functions. Thus, the convolution of *f* and *g* is given by Eq. (4.1).

$$(f^{*}g)(x,y) = \sum_{v=-\infty}^{\infty} \sum_{u=-\infty}^{\infty} f(u,v)g(x-u,y-v) \tag{4.1}$$

In this formula, *f* represents the original image, and *g* represents the convolution kernel. In practical applications, the kernel is defined over a finite set of points, according to the image processing task to be applied on the original image. Thus, considering a two-dimensional kernel of width $M = 2w + 1$ and height $N = 2h + 1$, the Eq. (4.1) becomes:

$$(f^{*}g)(x,y) = \sum_{v=-y-h}^{y+h} \sum_{u=x-w}^{x+w} f(u,v)g(x-u,y-v) \tag{4.2}$$

As an explanation for the equation representing the convolution, the reader can imagine *g* as one matrix "sliding" over the image *f* one pixel at a time. The result *f*g* in case of the pixel with coordinates *(x,y)* is the sum of the element-wise products of the two matrices. Figure 4.2 shows the convolution of a matrix and a kernel at *(x,y)* coordinate. In order to perform the complete convolution, the kernel is passed over each pixel of the original image. The result of the convolution represents a filtered version of the original image. There are various types of filtering performed through convolution with different kernels. These types of filters are commonly used for image denoising and edge detection.

Gauss filtering is a common first step in edge detection. The filtering is performed through a Gauss smoothing operator, or in other words, a 2-D convolution operator that is used to 'blur' images and remove noise. One such filter is called a Gauss, because the filter's kernel is a discrete approximation of the Gauss distribution. A circularly symmetric 2-D Gauss distribution has the form given by Eq. (4.3), where σ is the standard deviation of the Gauss distribution and the mean μ is *0*. Since the image is stored as a collection of discrete pixels we need to

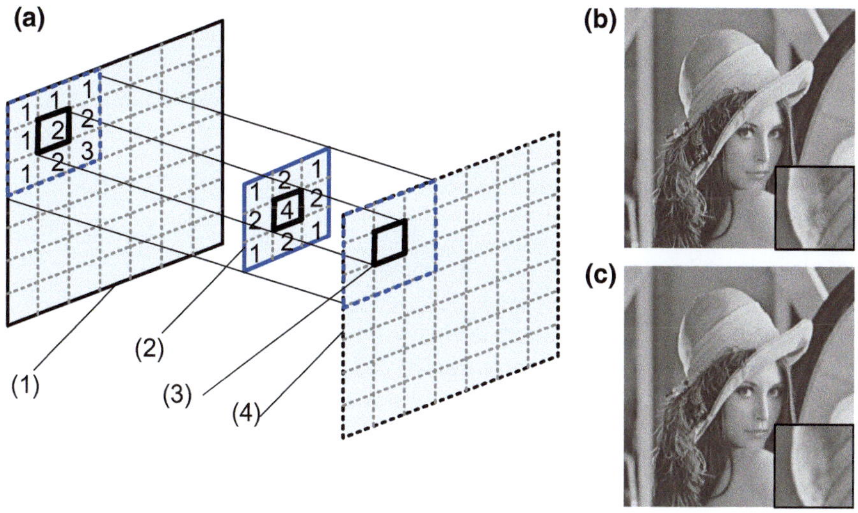

Fig. 4.2 **a** Convolution of the original image *I* and a gauss kernel at *(x,y)* coordinate; (*1*) the original image *I*, (*2*) Gauss kernel, (*3*) the location *(x,y)* from the filtered image, (*4*) the filtered image I_f; **b** The original image "Lena" together with hat detail in the right-down corner; **c** the image "Lena" filtered with a Gauss kernel

approximate the Gauss function, before performing the convolution. In practice, the Gauss kernel is approximated with *0* for more than about three standard deviations (3σ) from the mean μ. Thus, the kernel can be truncated at this point. Equation 4.4 shows a convolution kernel that approximates a Gauss. Figure 4.3a shows the plot of such 2-D kernel, where the similarity with the Gauss 2-D function from Fig. 4.3b is obvious.

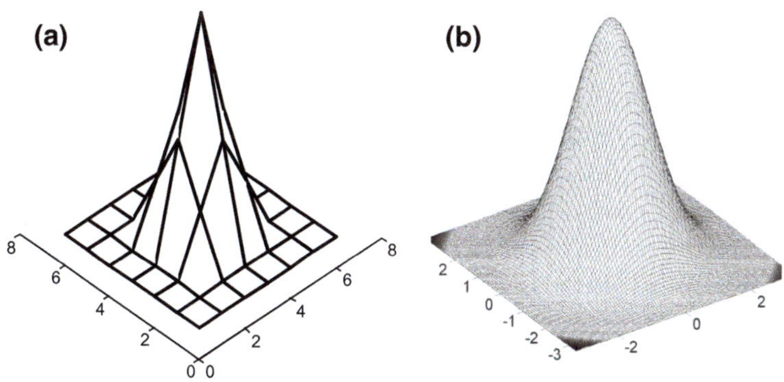

Fig. 4.3 **a** 2-D representation of a convolution kernel **b**, 2-D representation of the Gauss function

$$G(x,y) = \frac{1}{2\pi\sigma^2} e^{-\frac{x^2+y^2}{2\sigma^2}} \tag{4.3}$$

$$\frac{1}{16}\begin{pmatrix} 1 & 2 & 1 \\ 2 & 4 & 2 \\ 1 & 2 & 1 \end{pmatrix} \tag{4.4}$$

To conclude, one can say the convolution is used for the implementation of operators which have as output a linear combination of the intensities of pixels from the original image. Thus, each pixel of the output image is obtained by superimposing an $M \times N$ size window on the input image and calculating a linear combination of the pixels in the $M \times N$ window.

In case of a Gauss filter, which is generally used as a pre-processing technique in edge detection and other image processing algorithms, an example of a convolution kernel is described in Fig. 4.3a. Next paragraphs describe the implementation of a hardware implementation of image convolution. Through parameterization, the proposed architecture for convolution is set up to perform a Gauss filter and then used for the implementation of the Canny edge detectorin microarray images.

4.2.2 Hardware Implementation for Convolution

Once the appropriate kernel is chosen, convolution is applied as spatial filter to perform various image processing tasks. When computational efficiency is mandatory (real time applications), performing multiple convolution operations at the same time is a must. Moreover, for the same convolution operation, multiple computational steps are considered to be performed at once for efficient computation. Consequently, developing digital hardware architecture for the convolution operation represents a solution for efficient computation. Further on the approach proposed for the hardware implementation of the convolution is detailed. The VHDL code for the digital logic describing the proposed hardware architectures is also presented next.

The image to be filteredis first stored in a frame buffer memory. Let $M \times N$be the size of the convolution kernel and let *width* x *height* be the size of the image. The $M \times N$ window slides over the whole image and for each displacement thereof, the $M \times N$ pixels intensities values are taken for the calculation of a pixel from the output image (see Fig. 4.2). The constraints imposed by the memory make it impossible to acquire these $M \times N$ intensity values in one clock period; to overcome this limitation, a local caching operation is performed [7]. Pixels from *N-1* lines of the image together with *M* pixels from the *N*th line are stored using a shift register. This leads to the diagram shown in Fig. 4.4. Thus, in exchange for moving the M × N window on the surface of the image, the implementation delivers the input image, pixel by pixel, to the proposed shift register.

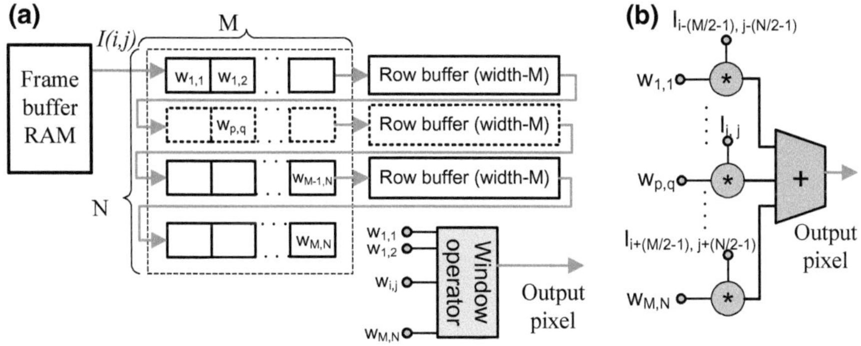

Fig. 4.4 Hardware architecture for image convolution composed of **a** the N shift registers for N-1 image lines and **b** the window operator

Once all *M(N−1)* + *M* registers cells are filled with pixel intensity values, the first pixel from the output image can be computed using the window operator (see Fig. 4.4b). A number of $N \times M$ multipliers are used together with an accumulator to calculate the output pixel value as a linear combination of pixels from the input image to be filtered. The input image pixels *I(i,j)* are delivered in a continuous manner to the hardware architecture; they are inserted in the shift register and the pixels from the output image *O(i,j)* are sequentially computed until last pixel from the input image *I(M,N)* is feed to the proposed architecture.

Example 4.1—Hardware architecture for Gauss filtering

The special case of the previous architecture which implements a Gauss filter using a 3 × 3 convolution kernel is designed using VHDL. Timing considerations are detailed only for the proposed hardware architecture for Gauss filtering. The methodology for developing the Gauss filter using VHDL language is summarized as follows: first, using the "black-box" approach, the inputs and outputs of the logic block for Gauss filter are defined; second, the behavioral description is detailed using the VHDL code; in order to check if the designed logic block has the expected functionality, a simulation is performed.

The corresponding "black-box" for the logical block designed to implement the Gauss convolution is shown in Fig. 4.5. The declaration of such logic block for Gauss convolution is detailed in the following VHDL code sequence:

```
———————————Entity declaration————————————————————————
  ENTITY Gauss_Filter IS
        PORT   (
                 START   : INstd_logic;
                 Reset   : INstd_logic;
                 clk     : INstd_logic;
                 enable  : OUTstd_logic;
                 Y       : INstd_logic_vector(7 downto 0);
```

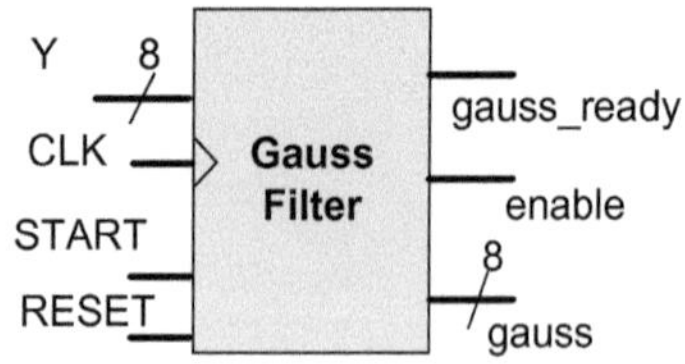

Fig. 4.5 Logic block for the Gauss filtering

```
                    gauss_ready       : OUTstd_logic;
                    gauss             : OUTstd_logic_vector (7 downto 0));
  END Gauss_Filter;
——————————End of entity declaration——————————————————————
```

The pixel intensity values are denoted by *Y* and its range between *0* and *255*. The *Start* signal marks the moment when the input has valid data and also the moment when processingthe input data *Y* (pixel intensity values) begins. The *enable* output pin marks the moment when the *gauss* output has valid data. The *gauss_ready* output marks the computation end for pixel intensity from the output image. It does not account if the computation delivers a valid data on *gauss* output or not. The VHDL behavioral description of the Gauss filter logic block is presented next:

```
—————————— Behavioral description of Gauss Filter ——————————
ARCHITECTURE Behavioral OF Gauss_Filter IS
—————————— constants defining the convolution kernel——————————
CONSTANT G_1379: unsigned (7 downto 0): = TO_UNSIGNED(1,8);
CONSTANT G_2468: unsigned (7 downto 0): = TO_UNSIGNED(2,8);
CONSTANT G_5: unsigned (7downto 0): = TO_UNSIGNED(4,8);
CONSTANT N: integer: = 10;                    --NxN size of the input image
CONSTANT pipe_level: integer: = 3;
——the registers "line_buffer" for the local caching of the input image——
SIGNAL line_buffer1: std_logic_vector(N*9-1 downto 0): = (Others => '0');
SIGNAL line_buffer2: std_logic_vector(N*9-1 downto 0): = (Others => '0');
SIGNAL line_buffer3: std_logic_vector(3*9-1 downto 0): = (Others => '0');
SIGNAL sum1,sum2,sum3,sum4,sum5,sum6,sum7,sum8,sum9,adder1,adder2:
UNSIGNED(23 downto 0);
SIGNAL j: integer range 102 downto 0: = 0;
SIGNAL done_pipe: std_logic_vector(pipe_level-1 downto 0): = (others => '0');
SIGNAL start_wre,temp: std_logic;
SIGNAL y_wre: std_logic_vector(7 downto 0);
```

```
BEGIN
gaussian: PROCESS (CLK, reset)
VARIABLE Q1,Q2,Q3: std_logic_vector(8 downto 0);
BEGIN
IF reset = ‘1’ THEN
        line_buffer1 <= (others => ‘0’);
        line_buffer2 <= (others => ‘0’);
        line_buffer3 <= (others => ‘0’);
ELSIF (clk’event and clk = ‘1’) THEN
        IF start_wre = ‘1’ THEN
                IF start_wre = ‘1’ THEN
                        Q1: = start_wre & y_wre;
                        Q2: = line_buffer1(8 downto 0);
                        Q3: = line_buffer2(8 downto 0);
                        ELSE
                        Q1: = “000000000”;Q2: = “000000000”;Q3: = “000000000”;
                END IF;
                line_buffer1 <= Q1 & line_buffer1(N*9-1 downto 9);
                line_buffer2 <= Q2 & line_buffer2(N*9-1 downto 9);
                line_buffer3 <= Q3 & line_buffer3(3*9-1 downto 9);
         END IF;

        IF (line_buffer3(8) = ‘1’) THEN
        sum1 <= UNSIGNED(X“00” & line_buffer1(N*9-2 downto N*9-9))*G_1379;
        sum2 <=UNSIGNED(X“00”& line_buffer1(N*9-11 downto N*9-18))*G_2468;
        sum3 <=UNSIGNED(X“00”& line_buffer1(N*9-20 downto N*9-27))*G_1379;
        sum4 <= UNSIGNED(X“00” & line_buffer2(N*9-2 downto N*9-9))*G_2468;
        sum5 <=UNSIGNED(X“00” & line_buffer2(N*9-11 downto N*9-18))*G_5;
        sum6 <=UNSIGNED(X“00” & line_buffer2(N*9-20 downto N*9-27))*G_2468;
        sum7 <=UNSIGNED(X“00” & line_buffer3(3*9-2 downto 3*9-9))*G_1379;
        sum8 <= UNSIGNED(X“00” & line_buffer3(2*9-2 downto 2*9-9))*G_2468;
        sum9 <=UNSIGNED(X“00” & line_buffer3(9-2 downto 9-9))*G_1379;
        adder1 <= sum1+sum2+sum3+sum4+sum5;
        adder2 <= adder1+sum6+sum7+sum8+sum9;
        ELSIF line_buffer2(N*9-1) = ‘1’ THEN
        adder2 <=(others => ‘0’);adder1 <=(others => ‘0’);sum1 <= (
others => ‘0’);sum2 <= (others => ‘0’);sum3 <= (others => ‘0’);sum4 <= (
others => ‘0’);sum5 <= (others => ‘0’);sum6 <= (others => ‘0’);sum7 <= (
others => ‘0’);sum8 <= (others => ‘0’);sum9 <= (others => ‘0’);
        END IF;
END IF;
END PROCESS gaussian;
```

```
——— process for the computational pipeline of the window operator ————
pipeline: PROCESS (CLK, reset)
BEGIN
IF reset = '1' THEN
                    done_pipe(0) <= '0';
                    done_pipe(1) <= '0';
                    done_pipe(2) <= '0';
                    gauss_ready  <= '0';
                    temp         <= '0';
ELSIF (clk'event and clk = '1') THEN
                    done_pipe(0) <= START_wre;
                    done_pipe(1) <= done_pipe(0);
                    done_pipe(2) <= done_pipe(1);
                    gauss_ready  <= done_pipe(2);
                    temp <=done_pipe(2);
END IF;
END PROCESS pipeline;
————— counter for the input pixel intensity values Y ——————
count: PROCESS (CLK, reset)
BEGIN
IF reset = '1' THEN
           j <=0;
ELSIF (clk'event and clk = '1') THEN
IF line_buffer2((N-1)*9-1) = '1' and (j < N*N) and done_pipe(2) = '1' THEN
             j <=j+1;
END IF;
END IF;
END PROCESS count;
—————————— concurrent assignments————————————————
Enable <= temp WHEN line_buffer2((N-1)*9-1) = '1';
gauss <= std_logic_vector(adder2(15 downto 8)) WHEN (line_buffer3(1*9-
1) = '1') ELSE
      "00000000";
start_wre <= start WHENj<(N-1)*N-1 or j>=N*N ELSE
                                        temp;
```

```
y_wre <= y WHENj < (N-1)*N-1  ELSE
                    "00000000";
END Behavioral;
```

The behavioral description of the Gauss filter logic block is composed of three processes and a section of concurrent assignments.

The first process called "*gauss*" is used to build up the shift registers for the local caching of image lines. For the ease of explanation, an image of *width x height* = *10* × *10* is considered for processing. The shift registers chain *line_buffer1*, *line_buffer2* and *line_buffer3* is filled sequentially with pixel intensity values. Moreover, each register cell contains also a supplementary bit which is set up to *"1"* logic value (*start_wre* signal) once the first intensity value is written in the registers. The *"1"* logic value is propagated to the last logic cell of the register chain. Once the aforementioned *"1"* logic value reaches the last cell of the registers chain, (*line_buffer3(8)* = *'1'*) the window operator computes the first output value *gauss*, which is given by the *adder2* operator and represents the first pixel from the output image.

The second process is a sequential one which generates the computational stages of the window operator. Depending on the kernel size, a different number of pipeline stages are needed. In case of a *M* × *N* = *3* × *3* kernel size, *9* multiplications are performed simultaneously and two accumulators are used to compute the result of the window operator. The number of pipeline stages is considered as defined by the constant "*pipe_level* = *3*". A shift register *done_pipe(2 downto 0)* is generated. For each pixel intensity delivered as input to the Gauss logic block a *"1"* logic value is written in the shift register cell *done_pipe(0)*. When *"1"* logic values reaches the *done_pipe(2)*cell, the output of the window operator is available on *gauss* output of the logic block.

The third process called "*count*" generates an *8-bits* counter which counts up to the number of pixels of the input image. For the ease of representation and simulation the input image is considered of size *10* × *10* pixels. The architecture can be easily parameterized to fit various image dimensions.

The concurrent assignments within the behavioral description generate the *enable* output which signalize when valid data is available at the *gauss* output. Also, the *gauss* output is assigned through a multiplexer as *0* value or as the result from the *adder2* accumulator.

Up to this point, the VHDL code is explained. Figure 4.6 details the simulation results in case of the architecture for the Gauss filter. Thus, the *Start* input marks the beginning of the computation for each pixel intensity value *Y*. The results are available on the *gauss* output when *"1"* logic value is found at single bit *gauss_ready* output. As compared with the *enable* output, the *gauss_ready* accounts also for the delay due to the local caching performed by the convolution architecture; thus it signalizes when first output data as part from the output image is available on the *gauss* output.

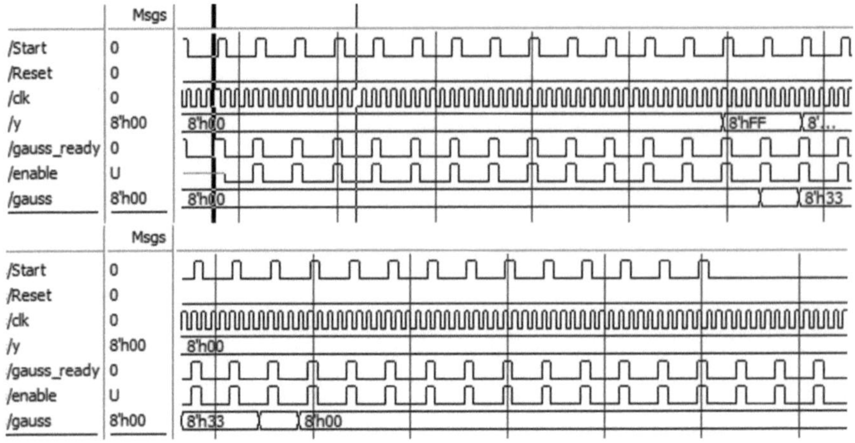

Fig. 4.6 Simulation of the logic block for the Gauss filtering

4.3 Hardware Architecture for the Canny Filter

The present chapter describes the Canny filter for edge detection together with its hardware implementation. The proposed implementation uses multiple instances of the architecture for convolution connected using a structural description which fulfills the steps of a classic Canny edge detection filter. A test-bench is designed for the simulation of the proposed architecture; moreover, the hardware architecture for edge detection is integrated as a co-processor within a microprocessor system to speed-up computation.

4.3.1 Canny Edge Detection

The edges are considered the locations of pixels where image brightness changes sharply, or, where high-intensity contrast is present in the image. The edges occur where the borders of objects are found; edge detection is widely used in image segmentation, when the main scope is to divide the image into areas corresponding to different objects. Image representation by its edges has the advantage that the amount of data is significantly reduced, while retaining most of the image information. The contours may be detected by applying a high-pass filter in the Fourier domain, or a convolution with a suitable kernel in the spatial domain. In practice, the detection is performed in the spatial domain, because there is less calculation and better results given.

The Canny algorithm is considered a "standard method" for edge detection and it is extensively used considering it delivers thin and pronounced contours. The Canny edge detector is a multistep process (Fig. 4.7). First it uses linear filtering

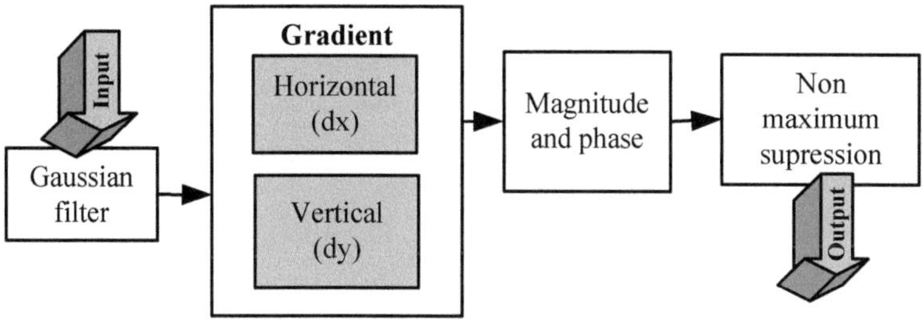

Fig. 4.7 Diagram for the contour detection using Canny filter

with a Gauss kernel to smooth the noise, and then calculates the force and the direction for each pixel location within the smoothed image. This is done by differentiating the image in two orthogonal directions (the second step) and calculating the magnitude and the direction of the gradient (the third step). In the fourth step of the process, the edge pixels are identified base on the pixel intensity gradient as the pixels that survive a process called non-maximum suppression. The diagram of the contour detection using the Canny filter is shown in Fig. 4.7.

The *first step* of the Canny edge detection, the Gauss filtering, was explained in detail in Sect. 4.2 together with a hardware implementation and its corresponding VHDL code.

After the image smoothing and noise elimination, the *second step*estimates the strength of the edges by computing the image gradient. Most edge detection methods work on the assumption that edges occur where a discontinuity appears in the pixel intensity. In most cases, edge detection operators can be considered gradient calculators. Considering a function *f* of one or more variables, the gradient vector field components are the partial derivatives of the *f* function, as denoted by Eq. 4.5.

$$\nabla f = \left(\frac{\partial f}{\partial x_1}, \ldots, \frac{\partial f}{\partial x_n} \right) \tag{4.5}$$

Let *I(x,y)* denote a digital image with *x, y* the spatial coordinates of image pixels. Considering images are discrete functions, for computingtheir gradient vector field we can use finite differences to approximate image derivatives on the two orthogonal directions, *x* and *y* respectively. It is well known that the derivatives are linear and shift-invariant; thus, the calculation of the gradient is most often achieved through a convolution. Various convolution kernels have been proposed to find edges, some of them are: Roberts, Prewitt and Sobel. In our case, Prewitt kernel is used; it is based on the simple idea that the difference between the lines and the difference between the columns are used for the horizontal gradient and for the vertical gradient respectively.

$$\frac{\partial I}{\partial x} \approx \frac{I(x+1,y) - I(x-1,y)}{2} \tag{4.6}$$

$$\frac{\partial I}{\partial y} \approx \frac{I(x,y+1) - I(x,y-1)}{2} \tag{4.7}$$

Regarding the hardware implementation for the vertical and horizontal derivatives, the architecture for convolution was used together with Prewitt convolution kernel and two simple signed adders to design a logic-block similar with the one from example 4.1. The designed logic block is used for the second step of the Canny filter. The corresponding VHDL code can be found in Appendix D.

The *third step* involves the computation of the magnitude and the direction of the gradient, denoted by P and Θ respectively. The absolute magnitude of the gradient $|P|$ is calculated by the square root of the sum of two squares: the image derivative on the horizontal direction (dx) and on the vertical direction (dy). To reduce the computational cost for the magnitude, P is approximated by the absolute value of the sum of the horizontal and vertical derivatives [see Eq. (4.8)].

$$|P| = \sqrt{dx^2 + dy^2} \approx |dx + dy| \tag{4.8}$$

The direction of the gradient (Θ) is computed according to Eq. (4.9):

$$\theta = \arctan\frac{dy}{dx} \tag{4.9}$$

The arctangent function requires at least fix-point precision for its calculation. There are CORDIC logic blocks available for trigonometric functions, but using them would significantly increase the hardware resource usage. Nevertheless, the range of integer values for *I(x,y)* together with 3 × 3 the kernel size used in our example, lead to the possibility to approximate both the magnitude and the direction of the gradient. The value and the sign of the gradient componentsare analyzed to calculate the direction of the gradient. Considering the pixel intensity value *I(x, y)*, the direction Θ of the gradient can be associated to one of the areas shown in Fig. 4.8, according to the derivatives sign and their absolute values $|dx|$ and $|dy|$.

The *fourth step* of the edge detection using Canny filter performs a selection of the pixels corresponding to edges within the image based on the computed gradient. This is done also based on image convolution, which allows that the values of the pixels under analysis to be compared with its neighboring pixels. The pixel that does not have a local maximum magnitude is eliminated. The comparison is made exclusively in the direction pointed by the gradient vector. For example, if the approximation of the gradient direction is between 0° and 45°, the magnitude of the gradient at the point $P_{x,y} = |dx| + |dy|$ is compared with the magnitude of the gradient at the points that are next to the current point, as it is shown in Fig. 4.9.

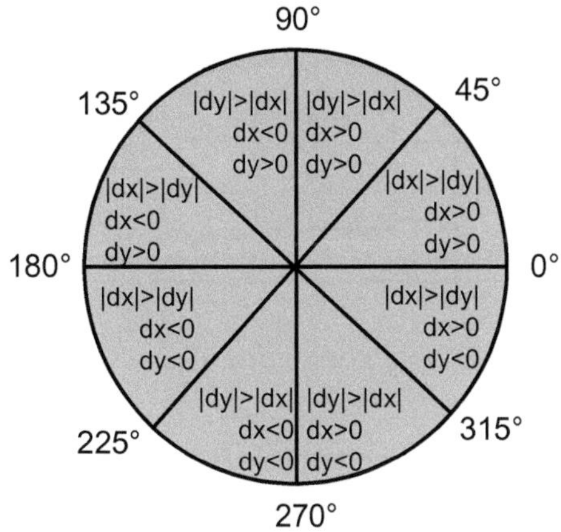

Fig. 4.8 Approximation of the gradient direction using *dx* and *dy* derivatives sign and their absolute values

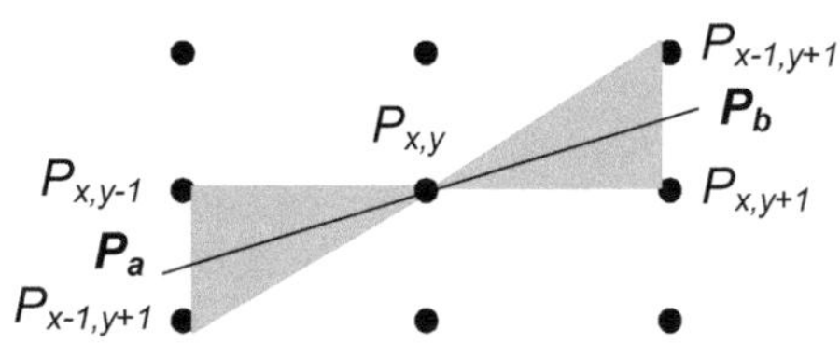

Fig. 4.9 Non-maximum suppression applied on pixel *(x,y)* based on gradient values *P*

$$P_a = \frac{P_{x,y-1} + P_{x-1,y+1}}{2} \tag{4.10}$$

$$P_b = \frac{P_{x-1,y+1} + P_{x,y+1}}{2} \tag{4.11}$$

The central pixel $P_{x,y}$ is considered as part of the edge if $P_{x,y} > P_a$ and $P_{x,y} > P_b$, where P_a and P_b are given by Eqs. (4.10) and (4.11) respectively. If the conditions are met, then the center pixel $P_{x,y}$ is removed (it is not considered a part of the edge).

4.3.2 Hardware Implementation of the Canny Edge Detector

Gauss filter for smoothing the image is implemented using the architecture detailed in example 4.1. Regarding hardware architectures for the *steps three* and *four*, the VHDL code for the corresponding logic blocks can be found in Annex C. The proposed architectures are similar to the one for Gauss filtering. Once the local caching of image lines is performed for the second step of the Canny filter, another two window operators are designed and used to compute the magnitude and

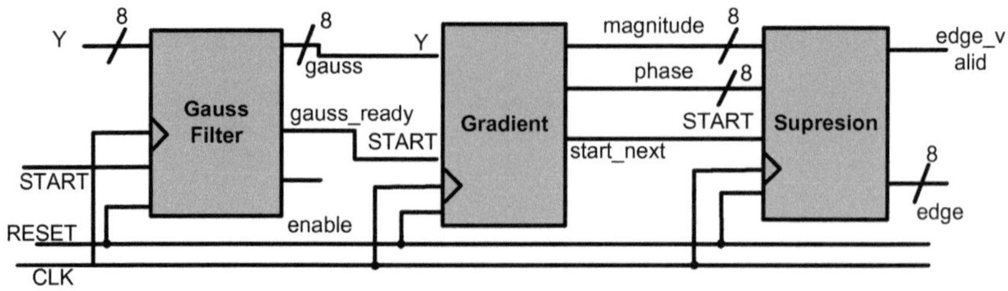

Fig. 4.10 Structural description of the Canny edge detection architecture

direction of the gradient P and Θ respectively (these two operators are included in the logic block called *gradient*). These values are input for another logic block called *suppression* associated to the non-maximum suppression, which also uses convolution architecture.

To sum up, three logic blocks are designed based on image convolution; the convolution is implemented with a local caching procedure performed by shift registers as in Fig. 4.4. The three logic blocks are: one for the Gauss filter (*Gauss Filter*), one for gradient computation (*Gradient*) and one for non-maximum suppression (*Suppression*) (Fig. 4.10).

Further on, an example with the VHDL structural description of the hardware architecture for the Canny filter is detailed.

Example 4.2—Structural description of Canny filter architecture
The declaration of the logic block, namely *Edge_Detector* for edge detection is presented in the next VHDL code sequence:

```
------------------ Entity declaration------------------------
    ENTITY Gauss_Filter IS
            PORT      (
                        CLk     : IN std_logic;
                        Start   : IN std_logic;
                        Reset   : IN std_logic;
                        y       : IN std_logic_vector (7 downto 0);
                        EDGE    : OUT std_logic_vector (8 downto 0);
                        EDGE_valid  : OUT std_logic;
                        send_READY  : OUT std_logic);
    END Gauss_Filter;
------------ End of entity declaration-----------------------
```

The proposed logic block has as inputs ports the following: *CLK* for the clock signal, *Start* for signaling when a pixel intensity value is delivered at the input port Y(pixel intensities values from the input image are sequentially delivered). The edge resulted image is delivered sequentially pixel by pixel at the*EDGE* output port (with a specific delay computed in the next Sect. 4.3.3); each pixel intensity from the output image is signalized by *'1'* logic at the *edge_valid* output. The *Edge_Detector* logic block is composed of three components interconnected as it is shown in Fig. 4.10. The *reset* port is used for initializing the shift registers with 0 before applying the hardware architecture for edge detection on a given image *I*.

```
————Structural description of the Canny edge detector ————
ARCHITECTURE Behavioral OF proc_chain IS
COMPONENT gauss_filter
           PORT(  Start    : INstd_logic;
                   Reset   : IN std_logic;
                    clk    : IN std_logic;
                    Y      : IN std_logic_vector(7 downto 0);
                    start_next: OUT std_logic;
                    gauss_ready : OUT std_logic;
                    gauss      : OUT std_logic_vector (7 downto 0));
END COMPONENT;
COMPONENTgradient
          PORT(   Start        : IN std_logic;
                   reset       : IN std_logic;
                   clk         : IN std_logic;
                   y           : IN std_logic_vector(7 downto 0);
                   start_next: OUT std_logic;
                   phase       : OUT std_logic_vector (1 downto 0);
                   magn        : OUT std_logic_vector (8 downto 0));
END COMPONENT;
COMPONENT suppression
         PORT(    Start        : IN std_logic;
                   reset       : IN std_logic;
                   clk         : IN std_logic;
                   data        : IN std_logic_vector(8 downto 0);
                   phase       : IN std_logic_vector (1 downto 0);
                  edge_valid : OUT std_logic;
                   edge        : OUT std_logic_vector (8 downto 0));
END COMPONENT;
———— wires declaration used to interconnect the three components————
CONSTANT PIPE_LEVEL : integer: = 3;
SIGNAL gauss_wre : std_logic_vector (7 downto 0);
SIGNAL magn_wre : std_logic_vector (8 downto 0);
SIGNAL phase_wre: std_logic_vector (1 downto 0);
SIGNAL gauss_ready_wre: std_logic;
SIGNAL start2: std_logic;
SIGNAL start3: std_logic;
```

```
—— instantiation of the three logic blocks and their interconnection——
BEGIN
uut1: Gauss_filter PORT MAP (CLK => CLK,
                START => START,
                       reset => reset,
                       y => y,
                       start_next => start2,
                       gauss_ready => gauss_ready_wre,
                       gauss => gauss_wre);
uut2: gradient PORT MAP (CLK => CLK,
                START =>START2,
                       reset => reset,
                       y => gauss_wre,
                       start_next => start3,
                       phase => phase_wre,
                       magn => magn_wre );
uut3: suppression PORT MAP (Start => START3,
                       CLK => CLK,
                       reset => reset,
                       data => magn_wre,
                       phase => phase_wre,
                       edge_valid => EDGE_VALID,
                       edge => EDGE);
     send_ready <=gauss_ready_wre;
END Behavioral;
```

As the VHDL code shows, three components are declared and instantiated, one for each computational step of the Canny edge detector. The signals declared within the aforementioned VHDL code are driven by wires which interconnect the three components. The result corresponds to three stages pipeline architecture for edge detection.

4.3.3 Timing Considerations for the Canny Edge Detection Architecture

Using the proposed approach for convolution, logic blocks are implemented for image smoothing, gradient computation and non-maximum suppression. The custom processing architectures detailed in Sects. 4.2 and 4.3 implement in a pipeline manner the independent processing steps of the Canny edge detector. In Fig. 4.11, the computational stages performed by the proposed architectures are denoted by (1)—*Gauss filter*, (2)—*Gradient* and (3)—*Suppression*. Let $t_0$mark the moment when the first pixel intensity value is delivered to the hardware architectures for processing. Depending on the size of the convolution kernel used in each of the processing architecture, there is a time delay up to the point the data is valid at the output of each of the architecture. Thus, in case of our hardware architectures where a 3 × 3 convolution kernels are used, three clock cycles ($3xT_{clk}$) delay is needed for data valid at the output. In other words, the time interval from t_0 to t_1 (computation end of the first architecture—*Gauss Filter*) is $3xT_{clk}$. Once the t_1 is reached, another pixel intensity value is delivered for processing to the *Gauss Filter* architecture (computational step (1')), whereas the architecture *Gradient* performs its calculation which also lasts for $3xTclk$ (t_1 to t_2) and it is denoted by computational step (2). In the same manner, pixel intensity values are delivered sequentially to the proposed architecture each with a delay computed as follows $\Delta t = t_1 - t_2 = 3xT_{clk}$; the resulted image is delivered at the output *edge* of the proposed Canny filter architecture at each Δt time interval. The output of the proposed architecture is visible also in Fig. 4.11. Thus, at the end of the computational steps denoted by (3), (3′) and (3″) three output pixels are delivered, corresponding to the inputs (1), (1′) and (1″) respectively. The delay between the input (1) and its corresponding output (3) is, obviously, $\Delta T_1 = 3x\Delta t$.

In case of the proposed hardware architecture for Canny filtering, another time delay ΔT_2 should be taken into account. This is due to the local caching procedure of the convolution operation used to perform the Gauss filtering. Thus, the registers *line_buffer1, line_buffer2* and *line_buffer3* (initialized with '0' at *reset*) have to be filled sequentially with pixel intensity values from the input image before delivering the first output. Moreover, another aspect should be taken into account: the output image size should be the same as the input image. Consequently, after $\Delta T_2 = width + 2$ pixel intensity values loaded in the shift registers (where *width,* represents the input image horizontal size) the first output pixel is available. The same delay is introduced by each of the three convolution operation. In this way, a $3x\Delta T_2$ time interval added to the initial ΔT_1, lead to a total delay path for the output image $\Delta T = 1 + 3(width + 2)\Delta t$.

The architecture for the Canny edge detection filter is simulated and the simulation results are discussed with respect to the timing considerations detailed previously. For the simulation to be performed, a test-bench is first created for the proposed architecture. The test-bench delivers pixel intensity values at the filter input and stores the resulted image. An example on how to construct the specific test-bench is presented next.

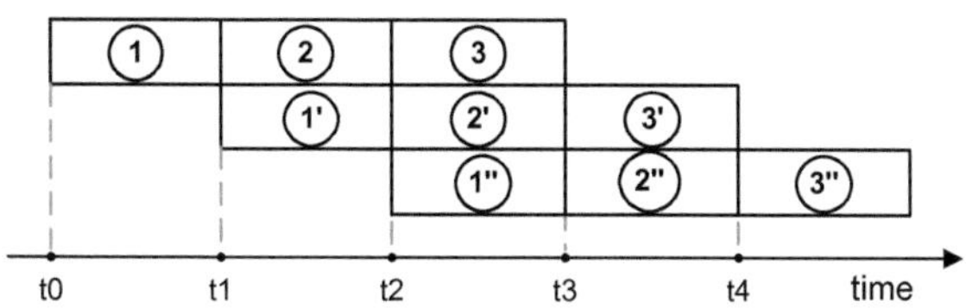

Fig. 4.11 Processing pipeline for the three independent steps of the canny edge detection filter: (*1*) Gauss filter, gradient computation and non-maximum suppression

Example 4.3—Test-bench for the Canny filter architecture
In order to test the functionality of the Canny filter architecture named *edge_detector*, the following VHDL code is used to build-up the test-bench. The test-bench is declared as any other logic block in VHDL and is named *test-bench_edge_detector*. Within the behavioral description of the *test-bench_edge_detector* architecture, the component *edge_detecor* is instantiated. Moreover, for each input and output of the architecture under test, the corresponding signals are declared. These signals will be used as wires to deliver input data to the architecture under test and also to monitor the outputs of the same *edge_detector* architecture.

```
ENTITY testbench_edge_detector IS
END testbench_edge_detector;
ARCHITECTURE behavior OF testbench_pchain IS
——Component Declaration for the Unit Under Test——————

COMPONENT edge_detector
        PORT(CLk : IN std_logic;
               Start : IN std_logic;
               Reset : IN std_logic;
               y : IN std_logic_vector(7 downto 0);
               EDGE : OUT std_logic_vector(8 downto 0);
               EDGE_valid : OUT std_logic;
               send_READY : OUT std_logic
              );
END COMPONENT;
——————— Inputs declaration ———————————————
SIGNAL CLk      : std_logic : = '0';
SIGNAL Start     : std_logic : = '0';
SIGNAL Reset    : std_logic : = '0';
SIGNAL y  : std_logic_vector(7 downto 0) : =  (others => '0');
———— Outputs declaration ———————————————
    SIGNAL EDGE : std_logic_vector(8 downto 0);
    SIGNAL EDGE_valid : std_logic;
    SIGNAL send_READY : std_logic;
CONSTANT clk_period : time : =  10 ns;
———————————— continued on the next code sequence—————————
```

Up to this point, the VHDL code is used for the instantiation of the unit under test, *edge_detector* architecture, and to declare the signals associated to the architecture inputs and outputs. Further on, the VHDL code uses different processes to deliver input data to the *edge_detector* and and to monitor *edge_detector* architecture outputs. The *clk_proc* processis used to generate the clock signal. The process *stim_proc* inserts stimulus to the architecture; sequentially reads the .txt file "*/microarrayspot.ini*" which contains the pixel intensity values of an image. The read pixel intensity values are delivered to the *Y* input of the *edge_detector* architecture. The next process called *writefile* stores the resulted Canny filtered image. The resulted image is stored pixel by pixel in the output.txt file "*/nC.ini*" by reading the output EDGE of the architecture and writing it using the *print* function.

```
———————continued from previous code sequnece————————————————
BEGIN
uut: proc_chain PORT MAP (
CLk => CLk,
          Start => Start,
          Reset => Reset,
         y => y,
         EDGE => EDGE,
         EDGE_valid => EDGE_valid,
         send_READY => send_READY);

clk_procs:PROCESS
BEGIN
        clk <= '0';
        wait for clk_period/2;
        clk <= '1';
        wait for clk_period/2;
END PROCESS;
——————————— process which inserts stimulus ————————————————
stim_proc: PROCESS
FILE Na:   TEXT open READ_MODE is "microarrayspot.ini";
FILE Nb:   TEXT open READ_MODE is "nB.ini";
VARIABLE LineNa: string(1 to 8),
VARIABLE LineNb: string(1 to 8);
BEGIN
    reset <= '0';
    WAIT for clk_period;
    reset <= '1';
    WAITFOR clk_period;
    reset <= '0';
    WAIT FOR 10*clk_period;
WHILE not endfile(Na) LOOP
          str_read(Na, LineNa);
          Y <= to_std_logic_vector(LineNa(1 to 8));
          start <= '1' ;
          WAIT FOR CLK_period;
          start <= '0' ;
          WAIT FOR 3*clk_period;
```

```
END LOOP;
WAIT;
END PROCESS;
——————————process used to store the output data in nC.ini file————
writefile: PROCESS
FILE Nc : TEXT open WRITE_MODE is "nC.ini";
VARIABLE LineNc     : string(1 to 9);
BEGIN
        WAIT UNTIL rising_edge (clk);
        IF EDGE_valid = '1' THEN
                    LineNc : = str(EDGE);
                    print(Nc, LineNc);
        END IF;
END PROCESS;
END;
```

The simulation results of the previous VHDL test-bench description is presented in Fig. 4.12. The test-bench performs the testing for the Canny edge detection architecture; it includes the reset of all logic blocks at the beginning. Further on, pixel intensity values are sent as inputs to the *Y* port of our Canny filter block. Each input of a pixel intensity value *Y* is marked by the *"1"* logic value on the *Start* input. For this example, a *width x height* = 10×10 input image *I* is considered for processing. Obviously, a 10×10 output image is found at the output *EDGE* with a delay consistent with the timing considerations detailed in Sect. 4.3.3. The end of the computation for each input value is marked by the *send_ready* value. This is not associated with pixel intensity value from the output image *O*, considering the delayed introduced by the local caching procedure for each computational step (Gauss filtering, gradient computation and non-maximum suppression). Thus, the total delay path of $\Delta T = 1 + 3(width + 2)\Delta t = 111\ T_{clk}$ (see Sect. 4.3.3—timing considerations) from the *Start* to the last pixel intensity from the output image is visible on the simulation from Fig. 4.12. The resulted 10×10 pixel intensity values are stored in the "*/nC.ini*" output file.

Running the simulation test-bench lead to generation of the "/nC.ini" file content. The file includes the resulted image after the architecture for Canny filter is applied. Figure 4.13 shows both the content for the initial image to be processed ("microarrayspot.ini") and the resulted Canny filtered image.

4.3.4 System-on-a-Chip (SoC) for Edge Detection

A SoC holds all the necessary hardware and software to form a complete system which serves a specific purpose, such as an image processing device. In order to

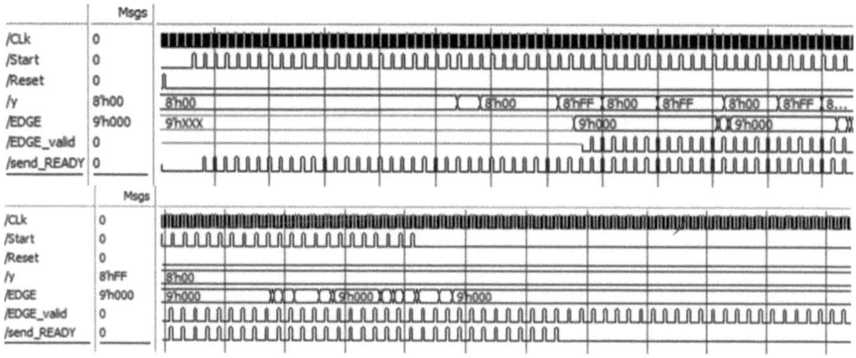

Fig. 4.12 Simulation results for the Canny edge detection hardware architecture: upper side the beginning of the simulation including the reset, lower-side the end of the simulation where *edge_valid* marks the pixel intensity values available for the output image

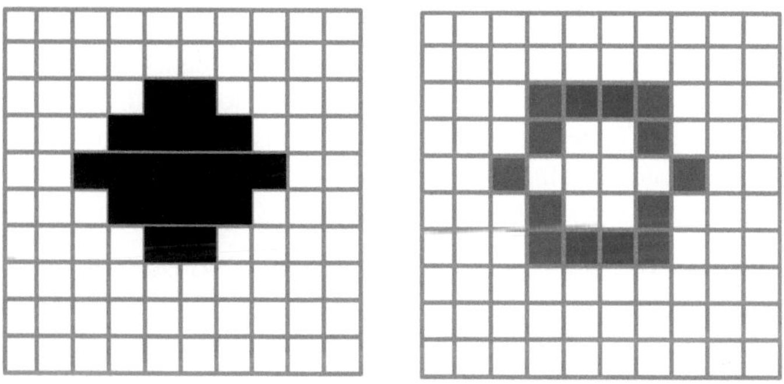

Fig. 4.13 Files contents with the initial grayscale image to be processed (**a**) and the result delivered by the proposed architecture for Canny filter (**b**)

take advantage of the proposed architecture for Canny edge detection, it has to be integrated within an image processing system. Commonly, these types of hardware architectures are used as custom hardware accelerators in microprocessor systems to speed-up computation. In our case, the Canny edge detection architecture is added to a Microblaze processor system-on-a-chip. Thus, as inspired by [8], the proposed logic block for edge detection is connected through the FSL (Fast Simple Link) data bus [9] to the Microblaze processor. The FSL protocol is used to delivered pixel intensities values to our proposed hardware architecture which is considered the slave device. The master device is the Microblaze processor which reads data from RAM and delivers data to the slave device; also the master device receives sequentially the results of the Canny edge detector filter. The write and read operation on the FSL bus are performed using the *getfsl* and *putfsl* c functions. FSL implements a point to point FIFO-based communication as shown in Fig. 4.14. FSL protocol involves two clock inputs*FSL_M_Clk* and *FSL_S_Clk* for master and slave

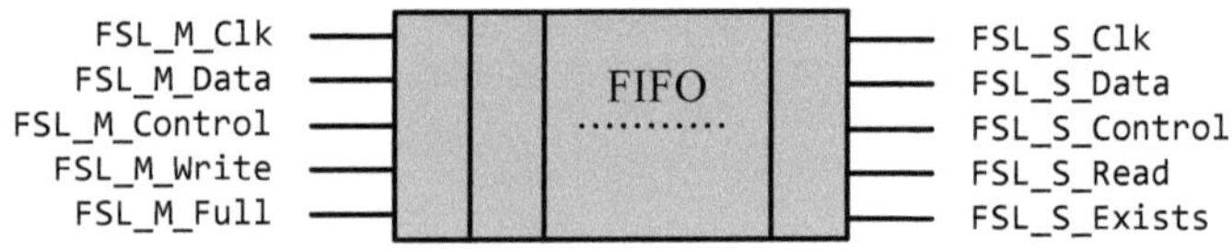

Fig. 4.14 FIFO-Based communication for the FSL bus

respectively, *FSL_S_Data* input port (for writing the pixel intensities to be processed into the FSL FIFO), *FSL_M_Data* output port (to read the resulted pixel intensity delivered by the Canny logic unit) to the FSL FIFO, *FSL_M_Write* and *FSL_S_Read* represent the control signal for read and write operation in and out of the FSL FIFO. *FSL_S_Exists* is a control signal which specifies if the FSL FIFO is empty or not.

As detailed in Fig. 4.15, the Canny edge detection logic block is defined by the inputs *Clk*, *Reset*, *Start* and *Y*. *Y* corresponds to the input pixel intensity values which are processed sequentially by the Canny logic block. The outputs of the Canny logic block are *edge* and *edge_valid*; *edge*correspond to pixel intensity values from the output image and *edge_valid* = *'1'* marks when valid data is available at the *edge* output. The Canny logic block is integrated within a Microblaze processor system using the FSL data bus. Taking into account the FSL protocol, a finite state machine (FSM) defined by *state_machine* entity is designed for the control of the proposed processing unit for Canny edge detector.

Example 4.4—VHDL description of the FSM for FSL control

The designed FSM has 4 states, *st_reset*, *st_wait*, *st_work* and *end_work*, and drives the Canny edge detector hardware implementation using the FSL data bus (see Fig. 4.16 for the FSM). The same input data as the one in Example 4.3 is considered for testing the architecture for edge detection: a 10×10 pixels size image is written in the FSL FIFO buffer using the *putfsl* function.

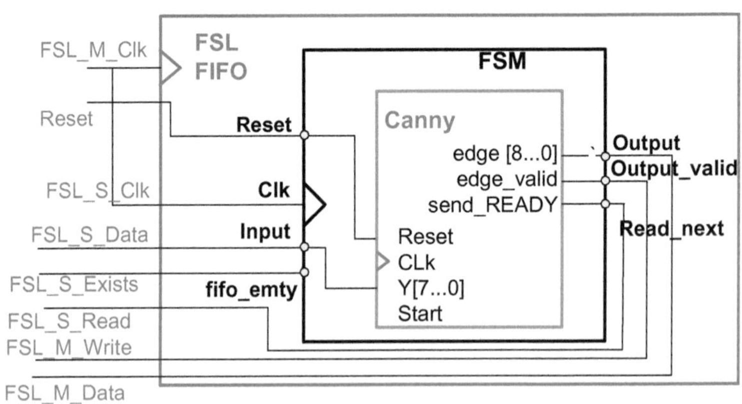

Fig. 4.15 Integration of the Canny edge detector processing unit into a microprocessor system through the FSL data bus using a specific FSM

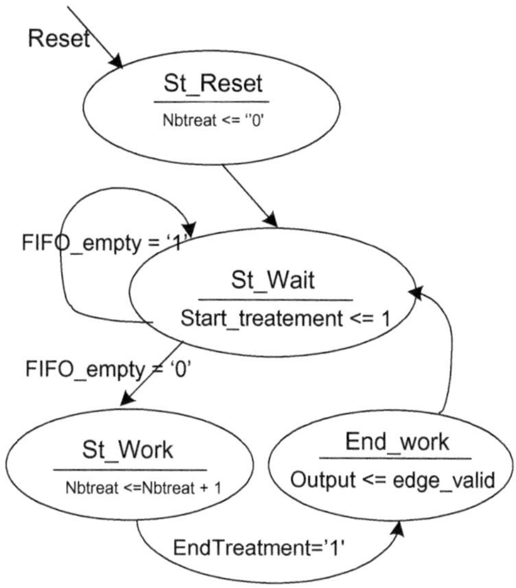

Fig. 4.16 FSM description for FSL control

The initial state *st_reset* initializes a counter *nbtreat* = *'0'*, for the number of pixels to be written in the FSL FIFO.

After reset, the next state is *st_wait* where, while FIFO is not empty (*FIFO_empty* = *'0'*), the pixel intensities are delivered to the *Y* port of the processing block through the *FSL_S_Data*; the *StartTreatment* output of the FSM is assigned to logic value '1'and it is connected to the input *Start* of the Canny filter architecture. Moreover, if *FIFO_empty* = *'0'* then the next state is *st_work.* Also a counter *nbtreat*is incremented to count the processed pixel intensities (*nbtreat* = *nbtreat* + *1*). The maximum value for the counter is 100 and it can be used to mark the end of delivering input data for processing (note: it is not used in this example).

In *St_work* state, the Canny architecture starts the processing, and through the output port "*send_ready*" the FSM goes in the state *"end_work"*. In other words, the read pixel intensities are processed, and when a result is available (*send_ ready* = *'1'*) the *EndTreatment* signalize the end of processing and the next state becomes *end_work.*

In the *end_work*state, the FSM delivers the control signal *FSL_S_read* to read the next pixel intensity from the FIFO to be processed, where the processing continues if *FIFO_empty* = *'0'* or the FSM waits for new values to be written in the FSL FIFO.

In the FSM description, there is also a concurrent assignment *output_ valid* <= *edge_valid* which is associated with the *FSM_M_write* which controls writing data from the *output* data bus to the FSL output FIFO. The *FSM_M_write* triggers the write operation of the data available on the *FSL_M_Data* bus into the FSL output FIFO.

The VHDL code sequence for the finite state machine *state_machine* which controls the delivery of data from *FSL_S_Data* bus to the Canny edge detector processing unit and reading the resulted data trough the *FSL_M_Data* bus is presented next.

```
————————VHDL description of the FSM for FSL control ————————
ENTITY State_Machine IS
        PORT( Start : in std_logic;
                Reset : in std_logic;
                clk     : in std_logic;
                input    : in std_logic_vector(31 downto 0);
                fifoempty : in std_logic;
                readNext : out std_logic;
                output: out std_logic_vector (31 downto 0);
                output_valid : out std_logic);
END ENTITY State_machine;

ARCHITECTURE fsmOF State_Machine IS
COMPONENT proc_chain IS
PORT (CLk                       : in std_logic;
        Start     : in std_logic;
        Reset     : in std_logic;
        y              : in std_logic_vector (7 downto 0);
        EDGE           : out std_logic_vector (8 downto 0);
        EDGE_valid     : out std_logic;
        send_READY: out std_logic);
END COMPONENT;



TYPE states IS (st_reset, st_wait, st_work,st_endwork);
SIGNAL current: states;
SIGNAL nbtreat : integer ;
SIGNAL startTreatment : std_logic;
SIGNAL endTreatment : std_logic;
BEGIN
procchain: proc_chain PORTMAP (
                CLk         => clk,
                Start       => startTreatment,
                Reset       => Reset,
                Y           => input (7 downto 0),
                EDGE        => output(8 downto 0),
                EDGE_valid          => output_valid,
                send_READY          => endTreatment);
```

```
PROCESS(clk)
BEGIN
IF rising_edge(clk) THEN
IF reset = '1' THEN
            current <= st_reset;
        nbtreat <= 0;
ELSE
CASE current IS
WHEN st_reset =>
                     current <= st_wait;
WHEN st_wait =>
IF fifoempty = '0' THEN
                           IF nbtreat = 100 THEN
                                    nbtreat <= 0;
                           ELSE
                                    nbtreat <=nbtreat + 1;
                           END IF;
                           current <= st_work;
                           startTreatment <= '1';
                     ELSE;
                           current <= st_wait;
                           startTreatement <= '0';
              END IF;
         WHEN st_work =>
IF endTreatment = '1' THEN
                              current <= st_work;
           ELSE
                              current <= st_wait;
  END IF;
                     startTreatment <= '0';



WHEN others => current <= st_reset;
            END CASE;
END IF;
      END IF;
END PROCESS;
Output_valid <= edge_valid;
Read_next <=startTreatment;
END ARCHITECTURE;
```

4.4 Canny Architecture Applied in Microarray Image Processing

Microarray image processing is chosen to exemplify the use of such hardware architecture in real-life applications. As described in the introductory section, microarray images include an increased number of microarray spots, up to 44 k, which leads to increased microarray image size. Microarray image processing uses image enhancement techniques together with PDE applied on vertical and horizontal image projections to estimate spot location. Once the spot location is established, segmentation is applied and, using border detection spot intensity extraction is performed and the level of expression for each gene is estimated. Thus, the differentially expressed genes are found by comparing the log odd ratios of the intensities from the two channel of the microarray image.

The layout of microarray images is suitable for processing multiple spots at once. Here, the benefits of spatial and temporal parallelism offered by FPGA technology come into play. Thus, on one side, the proposed approach for Canny edge detector allows performing fast computation for the edge detection in case of one microarray spot, and on the other side, multiple instances of the proposed Canny edge detection architecture can be used to processed multiple spots in parallel. Let the entire microarray image processing workflow be described by image enhancement using point-wise transform, profile computation and autocorrelation for spot location identification and Canny edge detection for segmentation. The levels of parallelization for the previously described image processing algorithms are discussed next.

Let *M*and *N*be the image dimensions. For image point-wise enhancement using logarithm transformation, due to the independent computation of logarithm for each pixelintensity value, multiple instances of an architecture similar to the one from Chap. 3 can be used. Considering p the number of instances of the logarithm computation units, the level of parallelization for image enhancement is $(MxN)/p$. For spot location identification using image profiles computation, PDE and autocorrelation, the level of parallelization is not a significant one, considering they are applied on image profiles. Thus, they are not considered for parallelization.

Once the spot locations are estimated, where k is the number of spots, the edge detection can be parallelized using the proposed architecture. Thus, the Canny edge detection hardware architecture is instantiated multiple times for parallel computation of microarray spots. Thus, for each spot, the hardware architecture of the Canny edge detector can be inferred. Nevertheless, the FPGAs resources are limited and, consequently, implementation constraints are involved regarding the number k of Canny edge detection hardware architectures. More details regarding the levels of parallelization can be found in [8]. The hardware resource usage in case of the XC5VlX110T FPGA chip for the implementation of the proposed Canny edge detection architecture are presented in the Table 4.1.

Table 4.1 Hardware resource usage for the implementation of 10 Canny edge detection hardware architecture for parallel microarray spot edge detection

Device utilisation summary (XC5VlX110T)			
Slice logic utilisation	Used	Available	Utilisation (%)
Number of slice registers	6125	69,120	8.8
Number of slice LUTs	8174	69,120	11.8
Number of bonded IOBs	129	640	20.2
Number of occupied slices	2328	17,280	13.4

The number of point-wise logarithm computation units is considered p = 100 for an $M \times N = 6100 \times 2160$ pixels Agilent image. The number of hardware architectures for parallel edge detection of microarray spots is k = 10. In Fig. 4.17, the abscissa axis represents different microarray images with different sizes given by the number of included microarray spots, whereas the ordinate axis represents the computational time for their processing. The computational time is evaluated using a personal computer and also the proposed application specific architectures implemented on XC5VlX110T Virtex5 FPGA chip [9].

The proposed application specific hardware architecture for edge detection in microarray images takes advantage of the parallel computation capabilities of the FPGA technology. The proposed implementation proved to be efficient with respect to the computational time (us). Thus, the experimental results based on algorithm parallelization show significant improvements of the proposed computation approach compared with a general purpose processor (PC) (Fig. 4.17). Over *80 ms* represents the total gain considering the computational time of the proposed architecture, as compared with a general purpose processor computational time. As expected, the FPGA technology is proved to be an efficient solution for an application-specific architecture for microarray image processing.

The proposed architecture for edge detection, connected as hardware accelerators to an FPGA-based processor system, is the first step towards an automated microarray image processing system. The main benefit of such system is the possibility to replace the workstation together with the software platform for microarray image processing with a system on a chip. Moreover, the proposed FPGA-based system can be easily integrated within the microarray scanner level. Due to the reduced computational time and cost, a large number of microarray analyses can be performed, compared with the existing computational tools.

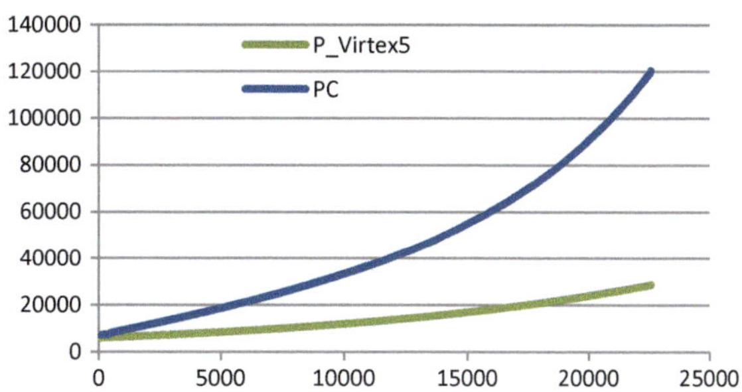

Fig. 4.17 Computational time for Canny edge detector filter implemented on a Virtex FPGA compared with the computational time needed by a general purpose processor for the same task

In the following chapters, application-specific hardware architecture for more complex automatic microarray image processing methods such us, partial differential equations (PDE)-based gridding or segmentation using circular Hough transform are detailed.

Appendix D

```
——————VHDL code for the non-maximum suppression processing step ————
library IEEE;
use IEEE.STD_LOGIC_1164.ALL;
use IEEE.NUMERIC_STD.ALL;

entity suppression is
   Port(   Start : in std_logic;
        reset : in std_logic;
        clk       : in std_logic;
        data      : in std_logic_vector(8 downto 0);
        phase : in std_logic_vector (1 downto 0);
        edge_valid: out std_logic;
        edge     : out std_logic_vector (8 downto 0));
end suppression;

architecture Behavioral of suppression is

constant N : integer : = 10;
CONSTANT PIPE_LEVEL : integer: = 3;

signal line_buffer1: std_logic_vector(N*12-1 downto 0): = (Others => '0');
signal line_buffer2 : std_logic_vector(N*12-1 downto 0): = (Others => '0');
signal line_buffer3  : std_logic_vector(3*12-1 downto 0): = (Others => '0');
signal Pa,Pb : UNSIGNED(9 downto 0);
signal j : integer range 102 downto 0: = 0;
signal done_pipe:
std_logic_vector (PIPE_LEVEL-1 downto 0): = (others  => '0');
signal start_wre,temp : std_logic;
signal data_wre : std_logic_vector(8 downto 0);
signal phase_wre : std_logic_vector(1 downto 0);

begin
edge_p : process (CLK, reset)
VARIABLE P1,P2,P3 : std_logic_vector(11 downto 0);
```

```
begin
   if reset = '1' then
        line_buffer1 <= (others => '0');
        line_buffer2 <= (others => '0');
        line_buffer3 <= (others => '0');
   elsif (clk'event and clk = '1') then
        if start_wre = '1' then
                 if start_wre = '1' then
                         p1 : = start_wre & phase_wre & data_wre;
                         p2 : = line_buffer1(11 downto 0);
                         p3 : = line_buffer2(11 downto 0);
                 else
                         p1: = "000000000000";
                         p2: = "000000000000";
                         p3: = "000000000000";
                 end if;
        line_buffer1 <= P1 & line_buffer1(N*12-1 downto 12);
        line_buffer2 <= P2 & line_buffer2(N*12-1 downto 12);
        line_buffer3 <= P3 & line_buffer3(3*12-1 downto 12);
        end if;

        if (line_buffer3(11) = '1') then
        if line_buffer2 ((N-1)*12-2 downto (N-1)*12-3) - "00" then
        Pb <= UNSIGNED('0' & line_buffer3
(3*12-4 downto 3*12-12)) + UNSIGNED('0' & line_buffer3
(2*12-4 downto 2*12-12));
        Pa <= UNSIGNED('0' & line_buffer1((N-2)*12-4 downto (N-2)
*12-12)) + UNSIGNED('0' & line_buffer1((N-1)*12-4 downto (N-1)*12-12));

         if (UNSIGNED (line_buffer2 ((N-1)*12-4 downto (N-1)*12-12)) >= Pa
(9 downto 1)) and (UNSIGNED (line_buffer2 ((N-1)*12-4 downto (N-1)
*12-12)) >= Pb(9 downto 1)) then
                edge <= line_buffer2 ((N-1)*12-4 downto (N-1)*12-12);
                         else
                                  edge <="000000000";
                         end if;
                end if;
                 if line_buffer2 ((N-1)*12-2 downto (N-1)
*12-3) = "01" then
         Pa <= UNSIGNED('0' & line_buffer3((3)*12-4 downto (3)
*12-12)) + UNSIGNED('0' & line_buffer2((N)*12-4 downto (N)*12-12));
         Pb <= UNSIGNED('0' & line_buffer1((N-2)*12-4 downto (N-2)
*12-12)) + UNSIGNED('0' & line_buffer2((N-2)*12-4 downto (N-2)*12-12));
         if (UNSIGNED (line_buffer2 ((N-1)*12-4 downto (N-1)
*12-12)) >= Pa(9 downto 1)) and (UNSIGNED (line_buffer2 ((N-1)
```

```
*12-4 downto (N-1)*12-12)) >= Pb(9 downto 1)) then
                edge <= line_buffer2 ((N-1)*12-4 downto (N-1)*12-12);
                        else
                                edge <="000000000";
                        end if;
                end if;
                 if line_buffer2 ((N-1)*12-2 downto (N-1)
*12-3) = "10" then
        Pa <= UNSIGNED('0' & line_buffer1((N)*12-4 downto (N)
*12-12)) + UNSIGNED('0' & line_buffer2(N*12-4 downto N*12-12));
        Pb <= UNSIGNED('0' & line_buffer3
(1*12-4 downto 1*12-12)) + UNSIGNED('0' & line_buffer2((N-2)
*12-4 downto (N-2)*12-12));
        if (UNSIGNED (line_buffer2 ((N-1)*12-4 downto (N-1)
*12-12)) >= Pa(9 downto 1)) and (UNSIGNED (line_buffer2 ((N-1)
*12-4 downto (N-1)*12-12)) >= Pb(9 downto 1)) then
                edge <=line_buffer2 ((N-1)*12-4 downto (N-1)*12-12);
                        else
                                edge <="000000000";
                        end if;
                end if;
                 if line_buffer2 ((N-1)*12-2 downto (N-1)
*12-3) = "11" then
        Pa <= UNSIGNED('0' & line_buffer3
(1*12-4 downto 1*12-12)) + UNSIGNED('0' & line_buffer3
(2*12-4 downto 2*12-12));
        Pb <= UNSIGNED('0' & line_buffer1
(N*12-4 downto N*12-12)) + UNSIGNED('0' & line_buffer1((N-1)
*12-4 downto (N-1)*12-12));
        if (UNSIGNED (line_buffer2 ((N-1)*12-4 downto (N-1)
*12-12)) >= Pa(9 downto 1)) and (UNSIGNED (line_buffer2 ((N-1)
*12-4 downto (N-1)*12-12)) >= Pb(9 downto 1)) then
                edge <= line_buffer2 ((N-1)*12-4 downto (N-1)*12-12);
                        else
                                edge <="000000000";
                        end if;
                end if;
        elsif line_buffer2(N*12-1) = '1' then
                edge <="000000000";
                pa <="0000000000";
                pb <="0000000000";

      end if;
```

```
        done_pipe(0)  <= START_wre;
        done_pipe(1)  <= done_pipe(0);
        done_pipe(2)  <= done_pipe(1);
        temp <= done_pipe(2);
end if;

end process edge_p;

debug: process (CLK, reset)
begin
if reset = '1' then
    j <=0;
elsif (clk'event and clk = '1') then
      if line_buffer2((N-1)*12-1) = '1' and j < N*N and done_pipe
(2) = '1' then
            j <=j + 1;
      end if;
      end if;
end process debug;

    edge_valid <= temp when line_buffer2((N-1)*12-1) = '1';
    start_wre  <= start when J < (N-1)*N-1 or j >=N*N else
                                                 temp;
    data_wre       <= data when J < (N-1)*N-1  else
                      "000000000";
     phase_wre  <= phase when j < (N-1)*N-1 and j >=2 else
                      "00";
    end Behavioral;
```

```
------------- Gradient computation filter (magnitude and phase)-----------
   library IEEE;
   use IEEE.STD_LOGIC_1164.ALL;
   use IEEE.NUMERIC_STD.ALL;

   entity Y_filter is
        Port( Start : in std_logic;
                          reset  : in std_logic;
                          clk      : in std_logic;
                          y        : in std_logic_vector(7 downto 0);
                          start_next: out std_logic;
                          phase : out std_logic_vector (1 downto 0);
                          magn     : out std_logic_vector (8 downto 0)
                          );
   end Y_filter;
```

```
   architecture Behavioral of Y_filter is

   constant N : integer: = 10;
   CONSTANT PIPE_LEVEL : integer: = 3;

   signal line_buffer1   : std_logic_vector
(N*9-1 downto 0): = (Others => '0');
   signal line_buffer2   : std_logic_vector
(N*9-1 downto 0): = (Others => '0');
   signal line_buffer3  : std_logic_vector
(3*9-1 downto 0): = (Others => '0');
   signal dY             : SIGNED (8 downto 0);
   signal dX             : SIGNED (8 downto 0);
   signal j : integer range 102 downto 0: = 0;
signaldone_pipe: std_logic_vector (PIPE_LEVEL-1 downto 0):
= (others => '0');
   signal start_wre,temp : std_logic;
   signal y_wre : std_logic_vector(7 downto 0);

   begin
   Y_p : process (CLK, reset)
   VARIABLE P1,P2,P3 : std_logic_vector(8 downto 0);

   begin
   if reset = '1' then
        line_buffer1 <= (others => '0');
        line_buffer2 <= (others => '0');
        line_buffer3 <= (others => '0');
   elsif (clk'event and clk = '1') then

        if start_wre = '1' then
                if start_wre = '1' then
                p1 : = start_wre & y_wre;
                p2 : = line_buffer1(8 downto 0);
                p3 : = line_buffer2(8 downto 0);
                else
                p1: = "000000000";
                p2: = "000000000";
                p3: = "000000000";
                end if;

                line_buffer1 <= P1 & line_buffer1(N*9-1 downto 9);
                line_buffer2 <= P2 & line_buffer2(N*9-1 downto 9);
                line_buffer3 <= P3 & line_buffer3(3*9-1 downto 9);
        end if;
```

```
        if (line_buffer3(8) = '1') then --and (j/= 20) and (j/= 21) and (j/
= 30) and (j/= 31)and (j/= 40) and (j/= 41)and (j/= 50) and (j/= 51)and (j/
= 60) and (j/= 61)and (j/= 70) and (j/= 71) and (j/= 80) and (j/= 81)then
        dY <= SIGNED ('0' & line_buffer1((N-1)*9-2 downto (N-1)
*9-9)) - SIGNED('0' & line_buffer3(2*9-2 downto 2*9-9));
        dX <= SIGNED ('0' & line_buffer2(N*9-2 downto N*9-9)) - SIGNED
('0' & line_buffer2((N-2)*9-2 downto (N-2)*9-9));
        if (abs(dY) > abs(dX) and (dY(8) = '0') and (dX(8) = '0')) or (abs
(dY) > abs(dX) and (dY(8) = '1') and (dX(8) = '1')) then
        phase <= "00";
        end if;
        if (abs(dX) > abs(dY) and (dY(8) = '0') and (dX(8) = '0')) or (abs
(dX) > abs(dY) and (dY(8) = '1') and (dX(8) = '1')) then
        phase <= "01";
        end if;
        if (abs(dX) > abs(dY) and (dY(8) = '1') and (dX(8) = '0')) or (abs
(dX) > abs(dY) and (dY(8) = '0') and (dX(8) = '1')) then
        phase <= "10";
        end if;
        if (abs(dY) > abs(dX) and (dY(8) = '1') and (dX(8) = '0')) or (abs
(dY) > abs(dX) and (dY(8) = '0') and (dX(8) = '1')) then
        phase <= "11";
        end if;
        elsif line_buffer2(N*9-1) = '1' then
                dY <= "000000000";
                dX <= "000000000";
        end if;

        done_pipe(0) <= START_wre;
        done_pipe(1) <= done_pipe(0);
        done_pipe(2) <= done_pipe(1);
        temp <=done_pipe(2);
    end if;

    end process Y_p;

    debug: process (CLK, reset)
    begin
    if reset = '1' then
        j <=0;
    elsif (clk'event and clk = '1') then
      --
In case of processing data from the previous gaussian block it writes
the magnitude results when there is valid data on the third buffer
    if line_buffer2((N-1)*9-1) = '1' and j < N*N and done_pipe(1) = '1' then
       j <=j + 1;
```

```
end if;
end if;
end process debug;

--
In case of processing data from the previous gaussian block it writes
the magnitude results when there is valid data on the third buffer

start_next <= temp when line_buffer2((N-1)*9-1) = '1';
magn <= std_logic_vector (unsigned(abs(dX)) + unsigned(abs
(dY))) WHEN (line_buffer3(1*9-1) = '1') ELSE
        "000000000";
start_wre <= start when J < (N-1)*N-1 or j >=N*N else
                                    temp;
y_wre          <= y when J < (N-1)*N-1  else
                 "00000000";
end Behavioral;
```

```
-------------Description of the Gaussian filter logic block------
library IEEE;
use IEEE.STD_LOGIC_1164.ALL;
use IEEE.NUMERIC_STD.ALL;
entity Gauss_Filter is
     Port( Start : in std_logic;
           Reset : in std_logic;
           Clk    : in std_logic;
           y      : in std_logic_vector(7 downto 0);
           gauss_ready: out std_logic;
           start_next: out std_logic;
           gauss   : out std_logic_vector (7 downto 0)
           );
end Gauss_Filter;

architecture Behavioral of Gauss_Filter is

CONSTANT G_1379: UNSIGNED (7 downto 0) := TO_UNSIGNED(21,8);
CONSTANT G_2468: UNSIGNED (7 downto 0) := TO_UNSIGNED(31,8);
CONSTANT G_5: UNSIGNED (7 downto 0) := TO_UNSIGNED(48,8);

constant N : integer:= 10;
CONSTANT PIPE_LEVEL : integer:= 3;

signal line_buffer1   : std_logic_vector
(N*9-1 downto 0):= (Others => '0');
signal line_buffer2   : std_logic_vector
(N*9-1 downto 0):= (Others => '0');
```

```
    signal line_buffer3   : std_logic_vector
(3*9-1 downto 0): = (Others => '0');

    signal sum1,sum2,sum3,sum4,sum5,sum6,sum7,sum8,sum9,adder1,
adder2 : UNSIGNED(23 downto 0);

    signal j : integer range 102 downto 0: = 0;
    signal done_pipe: std_logic_vector (PIPE_LEVEL-1 downto 0):
 = (others => '0');
    signal start_wre,temp : std_logic;
    signal y_wre : std_logic_vector(7 downto 0);
    begin

    gaussian : process (CLK, reset)

    VARIABLE Q1,Q2,Q3 : std_logic_vector(8 downto 0);

    begin
    if reset = '1' then
         line_buffer1 <= (others => '0');
         line_buffer2 <= (others => '0');
         line_buffer3 <= (others => '0');
    elsif (clk'event and clk = '1') then

         if start_wre = '1' then
                if start_wre = '1' then
                Q1 : = start_wre & y_wre;
                Q2 : = line_buffer1(8 downto 0);
                Q3 : = line_buffer2(8 downto 0);
                else
                Q1: = "000000000";
                Q2: = "000000000";
                Q3: = "000000000";
                end if;

                line_buffer1 <= Q1 & line_buffer1(N*9-1 downto 9);
                line_buffer2 <= Q2 & line_buffer2(N*9-1 downto 9);
                line_buffer3 <= Q3 & line_buffer3(3*9-1 downto 9);
         end if;

    if (line_buffer3(8) = '1') then --and (j/= 20) and (j/= 21)
and (j/= 30) and (j/= 31)and (j/= 40) and (j/= 41)and (j/= 50)
and (j/= 51)and (j/= 60) and (j/= 61)and (j/= 70) and (j/= 71)
and (j/= 80) and (j/= 81)then
         sum1 <= UNSIGNED(X"00" & line_buffer1(N*9-2 downto N*9-9))
*G_1379;
         sum2 <= UNSIGNED(X"00" & line_buffer1(N*9-11 downto N*9-18))
*G_2468;
```

```
        sum3 <= UNSIGNED(X"00" & line_buffer1(N*9-20 downto N*9-27))
*G_1379;
        sum4 <= UNSIGNED(X"00" & line_buffer2(N*9-2 downto N*9-9))
*G_2468;
        sum5 <= UNSIGNED(X"00" & line_buffer2(N*9-11 downto N*9-18))
*G_5;
        sum6  <= UNSIGNED(X"00" & line_buffer2(N*9-20 downto N*9-27))
*G_2468;
        sum7 <= UNSIGNED(X"00" & line_buffer3(3*9-2 downto 3*9-9))
*G_1379;
        sum8 <= UNSIGNED(X"00" & line_buffer3(2*9-2 downto 2*9-9))
*G_2468;
        sum9 <= UNSIGNED(X"00" & line_buffer3(9-2 downto 9-9))*G_1379;
        adder1 <= sum1 + sum2 + sum3 + sum4 + sum5;
        adder2 <= adder1 + sum6 + sum7 + sum8 + sum9;
        elsif line_buffer2(N*9-1) = '1' then
                        adder2 <=(others => '0');
                        adder1 <=(others => '0');
                        sum1 <= (others => '0');
                        sum2 <= (others => '0');
                        sum3 <= (others => '0');
                        sum4 <= (others => '0');
                        sum5 <= (others => '0');
                        sum6 <= (others => '0');
                        sum7 <= (others => '0');
                        sum8 <= (others => '0');
                        sum9 <= (others => '0');
    end if;
                --done_pipe(0) <= START_wre;
                --done_pipe(1) <= done_pipe(0);
                --done_pipe(2) <= done_pipe(1);
                --gauss_ready <= done_pipe(2);
                --temp <=done_pipe(2);

        end if;

    end process gaussian;

    end process gaussian;
    process (CLK, reset)
    begin
    if reset = '1' then
                done_pipe(0) <= '0';
                done_pipe(1) <= '0';
                done_pipe(2) <= '0';
```

```
            gauss_ready  <= '0';
            temp       <= '0';
elsif (clk'event and clk = '1') then
            done_pipe(0)  <= START_wre;
            done_pipe(1)  <= done_pipe(0);
            done_pipe(2)  <= done_pipe(1);
            gauss_ready <= done_pipe(2);
            temp <=done_pipe(2);
end if;
end process;

deb: process (CLK, reset)
begin
if reset = '1' then
    j <=0;
elsif (clk'event and clk = '1') then
      if line_buffer2((N-1)*9-1) = '1' and (j < N*N) and done_pipe
(2) = '1' then
      j <=j + 1;
     end if;
end if;

end process deb;

start_next <= temp when line_buffer2((N-1)*9-1) = '1';
gauss <= std_logic_vector(adder2(15 downto 8)) WHEN (line_buffer3
(1*9-1) = '1') ELSE
      "00000000";

--gauss <= std_logic_vector(adder2(15 downto 8));

start_wre <= start when J < (N-1)*N-1 or j >=N*N else
                                temp;

y_wre         <= y when J < (N-1)*N-1 else
              "00000000";

end Behavioral;
```

```
—————— Structural description of the Canny edge detector ——————
library IEEE;
use IEEE.STD_LOGIC_1164.ALL;
use IEEE.STD_LOGIC_ARITH.ALL;
use IEEE.STD_LOGIC_UNSIGNED.ALL;
```

```
entity proc_chain is
    port (
            CLk         : in std_logic;
            Start  : in std_logic;
            Reset : in std_logic;
            Y       : in std_logic_vector (7 downto 0);
            EDGE  : out std_logic_vector (8 downto 0);
            EDGE_valid      : out std_logic;
            send_READY: out std_logic
            );
end proc_chain;

architecture Behavioral of proc_chain is
component gauss_filter
        PORT(
                Start : in std_logic;
                Reset : in std_logic;
                Clk      : in std_logic;
                y         : in std_logic_vector(7 downto 0);
                start_next: out std_logic;     -- wired
                gauss_ready : out std_logic;
                gauss      : out std_logic_vector (7 downto 0));
end component;

component Y_filter
        PORT(
                Start : in std_logic;
                reset : in std_logic;
                clk      : in std_logic;
                y        : in std_logic_vector(7 downto 0);
                start_next: out std_logic;
                phase : out std_logic_vector (1 downto 0);
                magn       : out std_logic_vector (8 downto 0));
end component;

component suppression
Port(   Start : in std_logic;
     reset : in std_logic;
     clk      : in std_logic;
     data    : in std_logic_vector(8 downto 0);
     phase : in std_logic_vector (1 downto 0);
     edge_valid : out std_logic;
     edge      : out std_logic_vector (8 downto 0));
end component;
```

```
CONSTANT PIPE_LEVEL : integer: = 3;
signal gauss_wre : std_logic_vector (7 downto 0);
signal magn_wre : std_logic_vector (8 downto 0);
signal phase_wre: std_logic_vector (1 downto 0);
signal gauss_ready_wre: std_logic;
signal start2: std_logic;
signal start3: std_logic;

begin
uut1: Gauss_filter PORT MAP (
      CLK  => CLK,
      START  => START,
       reset  => reset,
       y  => y,
        start_next  => start2,
       gauss_ready  => gauss_ready_wre,
       gauss  => gauss_wre);

uut2: Y_filter PORT MAP (
          CLK  => CLK,
          START  =>START2,
          reset  => reset,
          y  => gauss_wre,
          start_next  => start3,
          phase  => phase_wre,
          magn  => magn_wre);

uut3: suppression PORT MAP (
        Start  => START3,
        CLK  => CLK,
        reset  => reset,
        data  => magn_wre,
        phase  => phase_wre,
        edge_valid  => EDGE_VALID,
        edge  => EDGE

send_ready <=gauss_ready_wre;
end Behavioral;
```

References

1. A.K. Whitchurch, *Gene Expression Microarray* (IEEE Potentials, Februarie 2002)
2. M. Schena, *Micropuce Biochip Technology* (Oxford: Oxford University Press, 1999)
3. D. Anastassion, *Genomic Signal Processing* (IEEE Signal Processing Magazine, Jully 2001)
4. V. Bolón-Canedo et al., A review of microarray datasets and applied feature selection methods. Inf. Sci. **282**(20), 111–135 (2014)
5. B. Belean, R. Terebes, A. Bot, Low-complexity PDE-based appraoch for microarray image addressing and segmentation, Med. Biol. Eng. Comput. **53**(2), 99–110 (2015)
6. J.F. Canny, A computational approach to edge detection. IEEE Trans. Pattern Anal. Mach. Intell. **8**(6), 679–698 (1986)
7. C.T. Johnston, D.G. Bailey, P. Lyons. A visual environment for real-time Imageprocessing in hardware. EURASIP J. Embed. Syst. 2006:072962 (2006)
8. B. Bot, S. Emerich, S. Martoiu and B. Belean, FSL-based hardware implementation for parallel computation of cDNA microarray image segmentation. Int. J. Adv. Comput. Sci. Appl. **6**(7), 2015
9. LogiCORE IP Fast SimplexLink (FSL) V20 Bus (v2.11c), Xilinx DataSheet, DS449 April 19, 2010

Chapter 5
Hardware Architectures for Iterative Algorithms Implementations

Equalities which contain functions of one or two variables and their partial derivatives are called partial differential equations (PDE). They are used to describe various phenomena such as heat propagation, sound or elasticity. They are also applied in image processing for smoothing and restoration purposes. Curves, surfaces or even the image itself are evolved according to PDE in such manner that the desired effect is obtained on the original image. Considering the discrete information found in digital images, these types of PDE filters are implemented using finite differences approximations of partial derivatives and iterative algorithms. The iterative nature of the PDE implementations represents a major disadvantage which leads to increased processing time in case of increased size images, or when several images need to be processed simultaneously. In this chapter two image processing applications which make use of PDE are described: *shock filters* applied in microarray image processing and *anisotropic diffusion* filter applied in satellite imagery. Their main disadvantage is increased computational time due to the iterative algorithms used for PDE implementation. In order to overcome it, application specific architectures are proposed for both the shock filter and the anisotropic diffusion filter.

5.1 Hardware Architecture for Shock Filters Applied in Microarray Image Processing

5.1.1 Partially Differential Equations in Image Processing

The effective use of PDEs in image processing can be credited to the fact that PDEs belong to one of the most important field of mathematical analysis and PDEs are strongly related to the physical world. Initially, PDE were used in physics and mechanics, e.g., the Maxwell equations in electromagnetics and Schrodinger

B. Belean, *Application-Specific Hardware Architecture Design with VHDL*, Signals and Communication Technology,
https://doi.org/10.1007/978-3-319-65025-8_5

equations in quantum mechanics. Nowadays, due to their power to describe and model dynamic processes (e.g. heat and sound propagation, fluid dynamics, etc.) they can be found in other fields such as biology, finance, artificial intelligence and computer science.

Images can be defined as a digital representation of a scene; consequently, its definition domain is represented by a discrete plane of points, called pixels. Therefore, a gray-level digital image will be equivalent to a function of spatial coordinates (x, y) and time t. PDE modeling of an image is achieved by means of a PDE function which has the luminance function and its partial derivatives as arguments.

Let $I_0(x,y)$ be the luminance intensity value associated with the pixel of coordinates (x, y) of the image I_0. The PDE filtered image represents a solution of the partial differential equation Eq. (5.1) at a given instant of time t, where F represents a function of the original image I_0 and its spatial partial derivatives.

$$\begin{cases} \frac{\partial I}{\partial t} = F(I(x,y,t)) \\ I(x,y,0) = I_0(x,y) \end{cases} \tag{5.1}$$

As an example, a reinterpretation of the gauss filtering can be done using PDE. Thus, PDE techniques consider the original image as initial state of a parabolic (diffusion-like) process, and extract the filtered versions from its temporal evolution according to the parabolic differential equation. The diffusion-like process (i.e. the heat propagation) is described by the Eq. 5.2, where α is a positive constant called diffusivity and ∇^2 is the Laplace operator.

$$\frac{\partial I}{\partial t} - \alpha \nabla^2 I = 0 \tag{5.2}$$

The main disadvantage of the isotropic diffusion performed by the Eq. 5.2 is that the filtering is performed in all directions, leading to contour smoothing. In the context of image processing applications such as feature extraction, any filter needs to preserve edge information. Thus, the proposed isotropic diffusion produces an unwanted effect by performing diffusion indiscriminately and altering the contours.

5.1.2 Shock Filters

By adding edge enhancement characteristics to the isotropic diffusion filtering leads to an image filtering method which performs the noise removing task whereas the edges are preserved. Such nonlinear filtering methods were proposed by by Osher and Rudin [1] for edge enhancement of blurry images and Perona and Malik [2] for anisotropic diffusion. The method proposed by Osher and Rudin is also known as *shock filter*.

Shock filters generally serve as an edge enhancing algorithm. Aiming blurry edge enhancement, Osher and Rudin proposed the first shock filter formulation based on a hyperbolic partial differential equation. The general one-dimensional (1D) shock filter model is described by Eq. (5.3), under the initial conditions $U(x,0) = U(x)$ and with the operator F fulfilling the following conditions: $F(0) = 0$ and $F(s) \times sign(s) \geq 0$. The U_x and U_{xx} represent respectively, the first and the second order derivatives. By choosing $F(s) = -sign(s)$, we obtain the classical shock filter Eq. (5.4).

$$\frac{\partial U}{\partial t} + F(U_{xx})|U_x| = 0 \tag{5.3}$$

$$U_t = -sign(U_{xx})|U_x| \tag{5.4}$$

A direct approach for numerically discretizing the classic shock filter is using the finite difference schemes for numerically approximating partial derivatives. The central difference approximation is performed using symmetrical approximation. For example, the approximation for the first order derivative of $U(x,t)$ is given by Eq. (5.5).

$$\frac{\partial U(x,t)}{\partial x} = \frac{U(x+k,t) - U(x-k,t)}{2|k|} \tag{5.5}$$

The aforementioned discretization scheme is not suitable since the shock filter model is an inverse of the diffusion equation, well known for its inherent instability. In overcoming this problem, Osher and Rudin propose the explicit discretization scheme detailed in Eq. (5.6), which preserves total variation and local extrema.

$$U(i)^{n+1} = U(i)^n - \Delta t \cdot |DU(i)^n| \cdot sign(D^2U(i)^n) \tag{5.6}$$

For a better understanding, in Eq. 5.6 the D and D^2 operators are described in Eq. (5.7):

$$\begin{aligned} DU(i)^n &= m(\Delta_+ U(i)^n, \Delta_- U(i)^n) \\ D^2U(i)^n &= (\Delta_+ \Delta_- U(i)^n) \end{aligned} \tag{5.7}$$

where $m(x,y)$ is the "minmod" function:

$$m(x,y) = [sign(x) + sign(y)] \cdot \min(|x|, |y|) \tag{5.8}$$

and $\Delta_\pm$ is:

$$\Delta_\pm = \pm(U(i \pm 1) - U(i)) \tag{5.9}$$

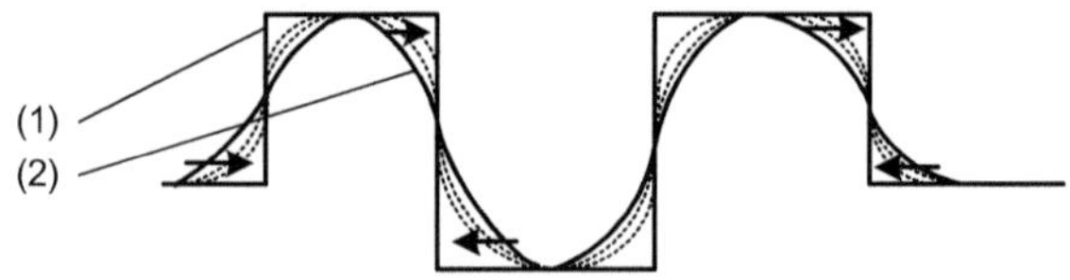

Fig. 5.1 1-Dimensional profile U_t evolution corresponding to the shock filter equation described in 5.4; (1) represents the original profile $U(x,0)$ whereas the resulted shock filtered profile is $U(x,n)$

In order to visualize the effect of the shock filters on an un-dimensional profile, the time evolution of the original profile U at different time scales t (see Eq. 5.4) is illustrated in Fig. 5.1. U_1 represents the original profile, whereas U_n represents the shock filtered profile after $t = 1 \ldots n$ iterations.

5.1.3 Shock Filter Application—Microarray Grid Alignment

Microarray technology is described in detail in the introductory section of this chapter. Its main objective is gene expression levels estimation which is performed as follows: gene specific probes from target and reference tissue samples labeled with two fluorescent markers are hybridized on the same glass slide. The specific probes together with their microscopic slide are called microarrays. The whole human genome which includes a number of approximately 44,000 genes can be printed on a microarray. Using double laser scanning, the microarray scanner scans the microarray slide and produces two microarray images, one for each fluorescent label. Each microarray image represents a collection of microarray spots, each spot determining the gene expression levels for a specific gene (e.g. in case of the target sample, its corresponding fluorescent label produces the microarray image *I*; within the image *I*, each spot represents the expression level of the represented gene). In order to determine the expression levels for each gene, the precise location of each spot within the microarray image has to be determined. The determination of microarray spots locations is considered the first image processing task within the microarray image processing workflow. This specific task is known as *grid alignment* or it can be simple called (a) gridding. An accurate determination of the gene expression level is a crucial step and involves also (b) *spot segmentation*, to classify pixels either as foreground, representing the DNA spots, or as background and (c) *extraction of intensity*, of each spot and its individual background. Results of the image analysis are the layout of the spot array, the spot sizes and shapes, the spot intensities (i.e., gene expression levels), and the background intensity values.

(a) Grid alignment assigns logic coordinates to each spot, or, in other words, determines the borderlines between adjacent rows or adjacent columns of spots, which leads to the determination of a rectangular area containing the microarray spot (see Fig. 5.2).

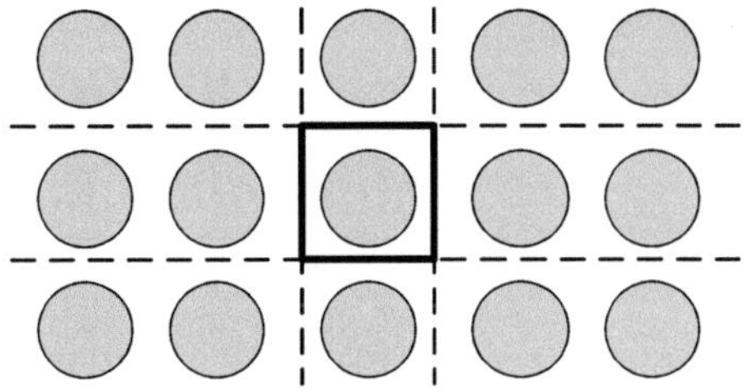

Fig. 5.2 The principle of grid alignment by delineating lines and columns of spots

Let us consider the input microarray image I defined by the two-dimensional array of intensities $I = (p_{u,v})$. The intensities $p_{u,v}$ are 16 bits integer with a dynamic range of $0 \leq p_{u,v} \leq 2^{16}-1$. A logarithmic transformation is applied for contrast enhancement, aiming to underline the weakly expressed spots. Consequently the I' image is obtained. On the resulted $I' = (p'_{x,y})$ image, grid alignment is performed by applying the uni-dimensional shock filters on the image profiles. The horizontal and vertical image profiles are computed as described by the Eqs. (5.10) and (5.11) respectively.

$$H(x) = \frac{1}{\dim_y} \sum_y P'_{x,y} \tag{5.10}$$

$$V(y) = \frac{1}{\dim_x} \sum_x P'_{x,y} \tag{5.11}$$

where dim_x and dim_y where the dimensions of the microarray image.
A shock filter was applied on the profiles H and V based on the partial differential scheme from Eq. (5.6). During shock filter iteration, the profiles converged to piece-wise constant functions.
The iteration produced discontinuities at positions $x_1, x_2, \ldots x_m$ and $y_1, y_2, \ldots, y_m$ of the inflection points of the horizontal intensity profle, H, and the vertical intensity profile V, respectively. The odd and even positions h_{2i}, h_{2i+1} underline the gap between two adjacent columns of spots. The center $x_c = (h_{2i} + h_{2i+1})/2$ is located centrally between adjacent maxima of the profile H, whereas $y_c = (y_{2i} + v_{2i+1})/2$, to separate rows of spots. The horizontal and vertical separation lines determination is lustrated in Fig. 5.3. The grid separates a spot from its neighbors and cuts the image into small rectangles, each of which contains a single spot. The resulted microarray grid in case of a microarray image is presented in Fig. 5.4.

(b) Spot segmentation classifies pixels to foreground and background, corresponding to the microarray spot or to its local background. The segmentation procedure starts with cutting the image I into sub-images $I_{row,i}$ and $I_{column,j}$. The sub-image $I_{row,i}$ is the horizontal slice of I that contains the i.th row of spots,

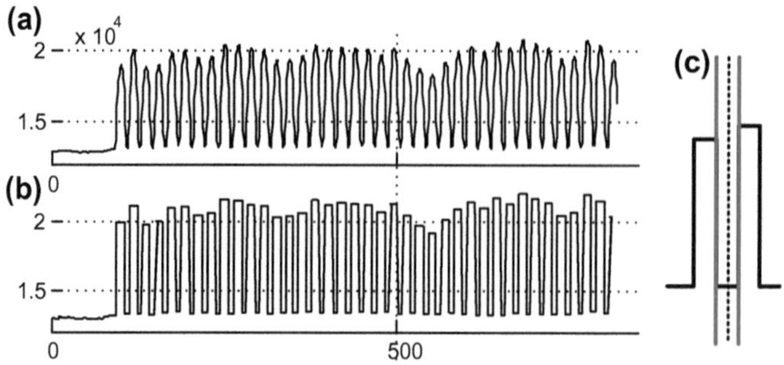

Fig. 5.3 **a** Microarray image horizontal profile, **b** The image profile after aplying shock filters, **c** The determination of line segments for separating columns of spots

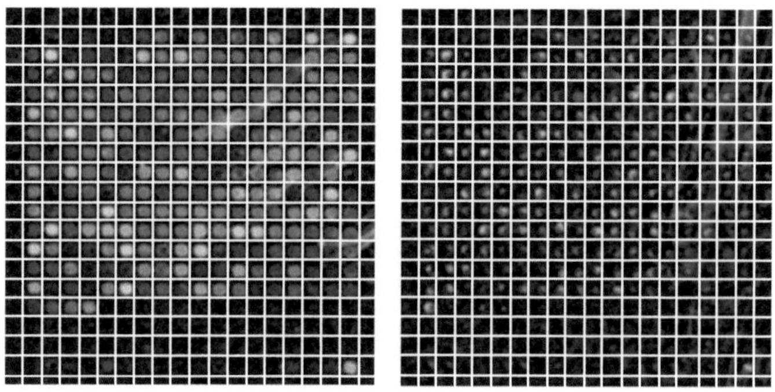

Fig. 5.4 Resulted grid obtained using shock filters for microaray spot detection on a microaray image sample

whereas $I_{column,j}$ contains j.th column of spots (see Fig. 5.5). Once the images containing rows and columns of spots are determined, horizontal image profiles and vertical image profiles are computed for $I_{row,i}$ and $I_{column,j}$ images. These profiles are once again iterated using the shock filter formalism. The resulted inflexion point represent the microarray spot horizontal and vertical margins, leading to the determination of a rectangular area which confines a micraoray spot as underlined in Fig. 5.5.

(c) Extraction of spot intensity is the last image processing step aiming for gene expression levels estimation. Pixel intensities, included within the rectangular area determined as specified in section (b), are assigned to one of the two groups: foreground (high value) and background (low value). The median intensity of the foreground pixels is associated with the gene expression levels. Foreground pixels determination can be performed using various spatial or distributional image processing techniques, e.g., k-means clustering, active contour. A detailed approach which makes use of k-means clustering as the segmentation procedure together with the corresponding results are presented in [3].

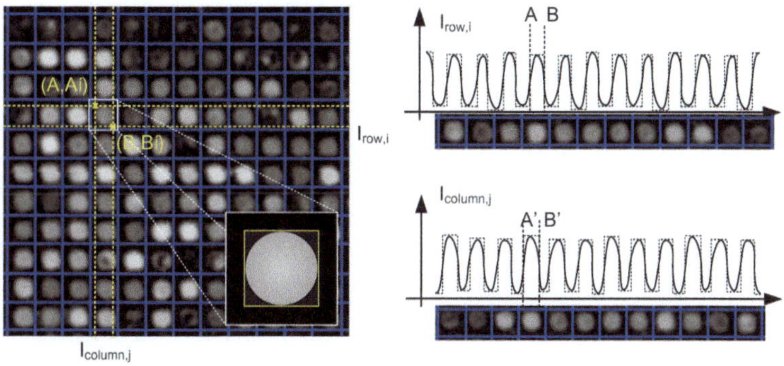

Fig. 5.5 Determination of the rectangular area corresponding to a given microarray spot

Notice that, considering the (a) and (b) image processing procedures, the main computational steps are the PDE-based profiles evolution according to the shock filter formalism. We estimate the computational complexity for our proposed approach for automatic grid alignment and segmentation, considering an $M \times N$ pixels size microarray image. The computational steps considering the horizontal profile computation and evolution are illustrated in Fig. 5.6a For the overall grid alignment procedure, the computational cost is given by the upper bound function $f(M, N) = 2MN\,s + 6p(M + N)s$, with s representing one computational step, and p denoting the number of iteration necessary for the profiles evolution. The order of

Algorithm: Shock filter based grid alignment — $f(M,N)$

Input: MxN size microarray image
$I(i,j)$ point-wise pixel intensity
Output: microarray horizontal grid alignment
U horizontal shock filtered profile

$(M \cdot N)s$:

for j ← 1 to N **do**
  for i ← 1 to M **do**
    $U(i) = U(i) + I(i,j)$
  end
end

$6(p \cdot M)s$:

for iter ← 1 to p **do**
  for i ← 1 to M **do**
    $U_{xx}(i) = U(i+1) + U(i-1) - 2U(i)$
    $\Delta\_U(i) = U(i) + U(i-1)$
    $\Delta_+U(i) = U(i+1) + U(i)$
    $fU(i) = U(i) - sign(U_{xx}(i)) \cdot \ldots$
    $\ldots \cdot \min(\Delta\_U(i), \Delta_+U(i)) \cdot \ldots$
    $\ldots \cdot [sign(\Delta_+U(i)) + \mathrm{sgn}(\Delta\_U(i))]$
  end
end

Algorithm: Shock filter based segmentation — $f(M,d)$

Input: Mxd size image - horizontal line of spots
$I_{row}(i,j)$ point-wise pixel intensity
Output: horizontal spot margins
U_r horizontal shock filtered profile

$(M \cdot d)s$:

for j ← 1 to d **do**
  for i ← 1 to M **do**
    $U_r(i) = U_r(i) + I_r(i,j)$
  end
end

$6(p \cdot M)s$:

for iter ← 1 to p **do**
  for i ← 1 to M **do**
    $U_{r\,xx}(i) = U_r(i+1) + U_r(i-1) - 2U_r(i)$
    $\Delta\_U_r(i) = U_r(i) + U_r(i-1)$
    $\Delta_+U_r(i) = U_r(i+1) + U_r(i)$
    $U_r(i) = U_r(i) - sign(U_{r\,xx}(i)) \cdot \ldots$
    $\ldots \cdot \min(\Delta\_U_r(i), \Delta_+U_r(i)) \cdot \ldots$
    $\ldots \cdot [sign(\Delta_+U_r(i)) + \mathrm{sgn}(\Delta\_U_r(i))]$
  end
end

Fig. 5.6 Computational steps for microarray grid alignment (*left*) and for the spot segmentation procedure (*right*)

growth for the computational cost is $O(f(M, N)) = 2MN + p(M + N)$ and represents the computational complexity of the proposed method.

The computational complexity of our PDE based segmentation procedure was estimated as follows. Let α and β represent the number of microarray spots on each line and columns, respectively, and d twice the average microarray spot diameter. The average width for a line or a column of spots is d. We computed for each spot line and spot column, the horizontal and vertical image profiles, respectively, with the total complexity of $2\alpha dM + 2\beta dN = 4MN$. Shock filters were further on applied on each of the determined profiles having a complexity of $p\alpha M + p\beta N$, where $p\alpha M$ represents p iterations performed on a number of α profiles (i.e. one profile for each line of spots), each profile having the size M. The total computational cost for both grid alignment and segmentation, led to the order of growth for the total computational cost for segmentation of $6MN + p(\alpha M + \beta N)$.

A comparison with state-of-the art microarray grid alignment and segmentation procedures is performed, considering the computational complexity. As reported in [mbec4], the autocorrelation based grid alignment has reduced computational complexity. Morphological operators for automatic microarray image addressing, have a computational complexity of O(2SeMN) where *Se* is the size in pixels (approx. 103) of the structural element for dilation and erosion. The computational complexity of the SVM-based approaches [5, 6] is O(MN(M + k)), one order of magnitude lower than the one associated to the genetic algorithm [7]. The parameter *k* represents the number of selected microarray spots to train the SVM. For the fully automatic microarray grid alignment performed using an optimal multilevel threshold (OMTG) approach [8], the reported computational complexity is O(tsN2), where ts denotes the threshold set size.

Regarding the segmentation procedure, the pixel clustering approach achieves lowest computational complexity by using a k-means clustering algorithm which has a time complexity of O(rkMN), [9]. Spot segmentation using mathematical morphology [10] has a computational cost $\gg$ SeMN, due to the morphological filtering by area opening with a structural element of the size of spot Se used to detect the initial markers for the watershed transform. The computational complexity of image segmentation using active contours can be reduced to n2MN, as reported in [11]. The n factor represents the size of a gauss kernel $\ll$ MN.

As denoted by Table 5.1, reduced computational complexity is achieved as compared to existing approaches. As illustrated by these results, our proposed approach has reduced computational complexity for both microarray image addressing and segmentation, being a strong candidate to be integrated in future software packages. Moreover, the proposed approaches for automatic grid alignment and segmentation are of high interest in case of application specific future devices for microarray image processing. By adding robust processing methods for gene expression microarray analysis and interpretation [12], future devices for medical applications which integrate the complete image processing pipeline can be developed [13].

Considering the benefits of the proposed image PDE-based image processing approach for grid alignment and segmentation (i.e. low-complexity and accuracy),

Table 5.1 Hardware resource usage for the anisotropic diffusion

Grid alignement	*Autocorelation*	*Mathematic morpjology*	*SVM*	*OMTG*	*PDE*
	O(M + N)	O(2S_eMN)	O(MN (M + k))	O ($t_s N^2$)	O(2MN + p (M + N))
Segmentation	*Pixel Clustering*	*Watershed transform*	*Active contours*	*PDE*	
	O(rkMN)	O(SeMN)	O (n^2MN)	$O(6MN + (p + d)(\alpha M + \beta N))$	

further on is proposed a hardware architecture for uni-dimensional shock filters. The proposed architecture aims to reduce the computational time, in spite of the iterative nature of the approach.

5.1.4 Hardware Architecture for Shock Filters

The aplication-specific hardware architecture automaticaly performs the grid alignment and delivers valuable information for the segmentation procedure. Thus, a rectangular area containg both the microarray spot (i.e. foreground information) and its corresponding local background is determined. The functionaity of the architecture is ilustrated in Fig. 5.7.

Image profiles are computed using the hardware architecture detailed in Chap. 1 and stored in Random access memories as ilustrated in Fig. 5.7. The computed profiles are stored within the shift register *"Buffer 1"* for concurential acces of

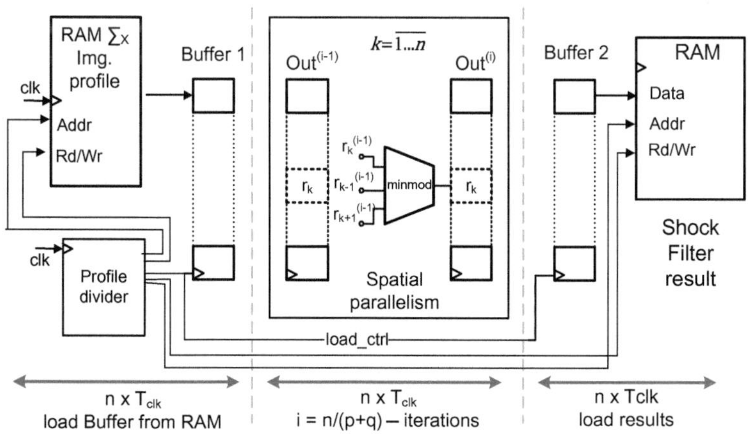

Fig. 5.7 Hardware architecture for shock filter implementation

profiles' values. Using the "*Profile divider*" logic block, the initial profile of length *L*, can be divided into succesive intervals of size *n*, for further processing. This division of the profile is performed especially in case of large image profiles (e.g. microarray image profiles). Timing consderations are also mentioned within the description. Thus, considering *n* the profile size, $n\ x\ T_{clk}$ clock periods are necessary to load the image profiles from RAM to the buffer register. Once the entire profile is loaded in "*Buffer 1*" register, a parallel load is performed and a copy of the profile is found in "Out^{i-1}" register. Whereas a new profile section is loaded into the "*Buffer 1*" register, the "Out^{i-1}" and "Out^{i}" make use of spatial parallelism to perform the computations corresponding to one shock filter iteration as given by Eqs. (5.12) and (5.13). The aforementioned computation procedure is specific for the shock filters and is known as the *minmod* function computation.

Loop $k = \overline{1 \ldots n}$

$$r_k^{(i)} <= r_k^{(i-1)} + dt \cdot \mathrm{sgn}(r_{k-}^{(i-1)} - r_{k+}^{(i-1)}) \cdot \min(r_{k-}^{(i-1)}, r_{k+}^{(i-1)}) \tag{5.12}$$

$$r_k^{(i-1)} <= r_k^{(i)} \tag{5.13}$$

End loop;

The logic block assigned to perform the *minmod* function computation is depicted in Fig. 5.8. The logic block delivers the resulted r_k^i value within $p = 3x\ T_{clk}$. A number of *n* instances of the proposed architecture are used for parallel computation of the r_k^i values. Another $q = 1\ x\ T_{clk}$ is used to create the *Buffer 2* register which contains a copy of the $Out^{(i)}$ register for further processing.

In order to have a fully pipelined architecture, the number of iterations *i* for the shock filter is chosen as $i = n/(p + q)$. A number of $i = 50$ iterations was empirically determined in such manner that the resulted evolved profile underlines the inflexion points.

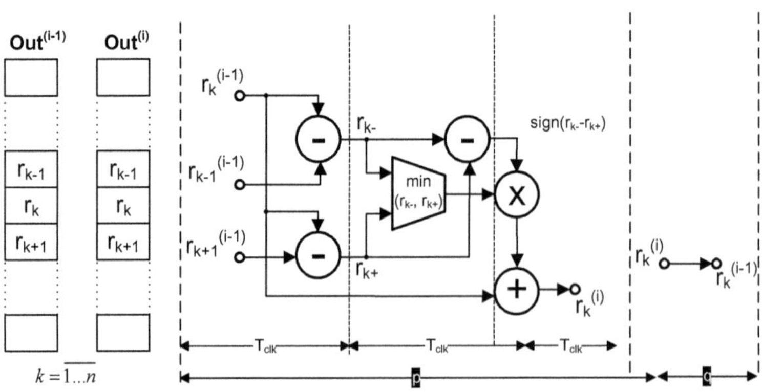

Fig. 5.8 Hardware architecture for *minmod* function implementation

Table 5.2 Hardware resource usage for uni-dimensional shock filters

No. of shift register cells (16 bits size)	No. of multipliers (16 bits)	No. of LUT based ROM (256 kb)	No. of adders (16 bits)	Comparators (16 bits)
640	128	2	512	128

5.1.5 Timing Considerations

Once the *minmod* function computation is described, we can go back to the overall architecture for the shock filter profile computation (see Fig. 5.8) and clearly describe the timing consideration. The architecture is fully pipelined, considering the following stages: *stage 1* loads a section of *n* values from the original image into the *Buffer 1* register; *stage 2* performs $i = n/(p + q)$ iterations on the image profile section of size *n*; *stage 3* stores the resulted shock filtered profile in the RAM memory. With an initial delay of $\Delta t = 2\ x\ nT_{clk}$, the shock filtered 1D profile values are sequentially stored into the output *RAM* memory.

Table 5.2 summarizes the hardware resource usage for the proposed architecture. The computational time needed for performing automatic grid alignment and shock filter based segmentation was estimated in case of Intel Dual core T2370 processor, with a 1.73 GHz clock frequency, 2 GB RAM and Virtex5 platform. In case of the general purpose processor C code and *clock()* function were used for time measurement. The hardware architecture for shock filter implementation was applied on images containing different number of spots. Let *X* axis referring to the dimension of microarray image specified by the number of microarray spots enclosed and *Y* axis represents the processing time in microseconds (us). Thus, in Fig. 5.9 can be observed that, in spite of higher performances provided by the general purpose processor, the application-specific architectures bring up better results, considering the computational time used for grid alignment and PDE-based segmentation.

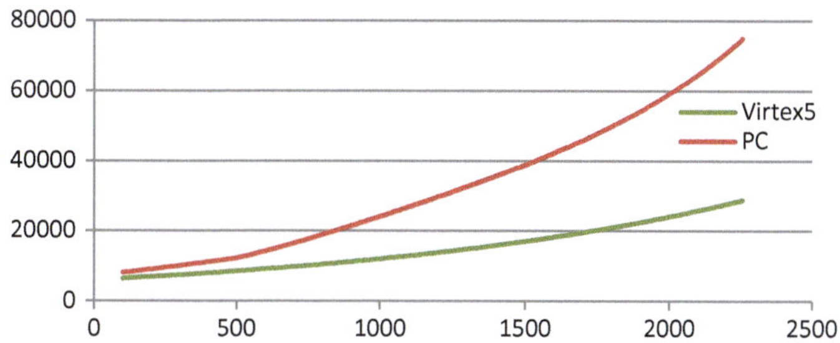

Fig. 5.9 Computational time for grid alignment and segmentation using a general purpose processor and Virtex5 FPGA

5.2 Hardware Architecture for Anisotropic Diffusion Applied in Satellite Imagery

5.2.1 Introduction to Satellite Imagery

Satellite and aerial images have many useful applications in fields such as agriculture, forestry, geology, meteorology, mass-media or regional planning. Software developers incorporate satellite imagery into stand-alone platforms or even hand held devices. Thus, satellite imagery is combined with GPS and electronic maps creating geographic information systems which lead to the real time localization of various targets (objectives) within specific areas. The most popular example of this is the Google Earth application, which recently made commercial satellite imagery freely available to almost anyone on the planet. Some of the industries also benefit from Google Earth services. The presence of satellite and aerial images on the Internet provides an opportunity for exponential growth and various benefits for both private and public sectors.

Satellite imagery is commonly compared with classic imagery. Regular photographic cameras store information about the portion of light called "light visible". The information is stored as images, and represents the interaction of the visible light with the area under analysis. The human eye quantizes the interaction of visible light with different targets in a similar manner in order to provide us visual images. Using various sensors, satellites are sensitive to other parts of the electromagnetic spectrum, such as infrared, ultraviolet or microwaves. The sensors measure "light" (i.e. electromagnetic radiation emitted by earth) and use computer programs for creating images. An in depth view of how satellites acquire images is as follows: an energy source emits a radiation which interacts with the atmosphere and then reaches the target; the target, depending on its characteristics, reflects the radiation to the sensor which registers it and creates a digital image. Image data are recorded at different spatial, spectral and temporal resolutions. The interpretation of the image leads to the characterization of the target and to discovery of the desired aspects related to the target under analysis.

There are various applications of satellite image data, from environmental monitoring (e.g. indirect solar radiation measurements [14], climate change measurements [15] to natural disaster monitoring and mitigation [16]). For satellite image analysis and interpretation, various image processing techniques are available, from simple convolution for edge detection to complex partial differential equations or neural networks for the same edge detection task. Partial differential equations (PDE) have various applications in image processing and computer vision.

The increased number of high resolution satellites launched into orbit and the increased number of applications which use satellite images creates the premises for a large amount of high-resolution image data to be processed. Perona and Malik filter is commonly used in image processing application and it is known as a computationally expensive approach due to its iterative PDE-based implementation. A challenge in recent research is to develop high performance computing models for satellite image processing [17, 18]. Parallel computing principles such as LUT (look-up-tables) were used for high performance computing in solar surface irradiance estimation [19]. Further on we briefly described the Perona and Malik filter formulation and applications together with FPGA-based application-specific hardware architecture for fast and efficient Perona and Malik filter implementation. The benefits of the proposed architecture regarding computational time, together with the results of anisotropic diffusion in case of satellite images containing solar surface irradiance levels are detailed. Satellite images were chosen for exemplification of the anisotropic diffusion and its hardware implementation, due to their intensive use and increased size.

5.2.2 Perona and Malik Filter Formulation

PDEs are used to describe a wide variety of phenomena such as sound, heat, electrodynamics, fluid flow, elasticity, or quantum mechanics. In image processing they are mainly used for smoothing and restoration purposes. Their main advantages are the reinterpretation of several classical methods - such as gauss convolution, median filtering, dilation or erosion. This understanding has also led to the discovery of new methods for shape simplification, structure preserving filtering, and enhancement of different structure types. Aiming edge enhancement, typical PDE techniques consider the original image as initial state of a parabolic (diffusion-like) process, and extract the filtered versions from its temporal evolution according to the parabolic differential equation.

The physical idea behind diffusion processes is that it describes the distribution of intensity (e.g. variation in temperature or variation of luminance information—pixel intensity in case of image processing) in a given region over time. The diffusion is known as a physical process that equilibrates concentration differences (i.e. luminance information in image processing) without creating or destroying mass. The mathematical formulation is given by the following equilibrium property [2]:

$$j = -D \cdot \nabla u \tag{5.14}$$

D is a diffusion tensor represented by a positive symmetric matrix, which establishes the relation between the concentration gradient ∇u and a flux j which aims to compensate for this gradient. In case j and ∇u are parallel, the diffusion is called isotropic. The property of the diffusion of not to destroy mass/information is expressed by the continuity Eq. (5.15).

$$\partial_t u = -div\, j \tag{5.15}$$

Considering all of the above, the diffusion equation is given by the time dependent Eq. (5.16). Considering the diffusion tensor, if it depends on the image ut which evolves in the time domain, the diffusion is called non-linear. Moreover, if the diffusion tensor D is constant over the whole image domain, the diffusion is considered homogeneous or isotropic, whereas a space-dependent diffusion tensor is called inhomogeneous or anisotropic.

$$\partial_t u = div(D \cdot \nabla u) \tag{5.16}$$

Anisotropic diffusion for edge enhancement.

Perona and Malik [2] propose a nonlinear diffusion method which preserves the edge information. For avoiding the blurring at edge locations, the inhomogeneous diffusion process reduces the diffusivity at those locations which have a larger likelihood to be edges. The probability for a specific area to be edge is denoted by $|\nabla u|^2$. Consequently, the Perona–Malik equation is (5.17).

$$\partial_t u = div(g(|\nabla u|^2) \cdot \nabla u) \tag{5.17}$$

where the diffusivity function g is to be anisotropic, such that, along some directions $g(|\nabla u|^2) = g(s) \gg 1$ (strong diffusion) while along other directions $g(s) \ll 1$ (weak smoothing).

The use of the Eq. (5.17) is motivated by the physical fact that, under $g \equiv 1$ it describes the heat propagation. The solution of the heat equation is a convolution of the initial value $u(x, 0) = f(x)$ with a Gauss distribution function. Since the latter is decreasing fast (both in the coordinate and the frequency spaces), the fast oscillations are cut out, hence (10) acts as an effective low pass filter.

The digital image processing based on the Eq. (5.17) is motivated by the following considerations. There is a bounded domain of the digital image, $\Omega \subset R^n$ ($n = 2, 3$) of boundary $\partial\Omega$ of class C^1. The mapping u: $\Omega \rightarrow [0, 1]$ then achieves the correspondence from Ω to the gray level distribution (GLD) of a noisy image. The numerical investigation of the time evolution of the GLD, performed through an iterative approach, results in successive instances attempting at solving the filtering tasks.

There are two contradictory features of the gauss smoothing associated to the heat propagation: efficient noise filtering and image blurring, which results in quick loss of essential information contained in the original image. In order to take advantage of both features, a well posed condition problem arises. The flux function $\Phi(s) = s \cdot g(s^2) > 0$ for $s \in (0, +\infty)$ is wanted to vary smoothly with s and to have a maximum on $(0, +\infty)$ at some characteristic value $s_0 = \lambda > 0$. The diffusivity function $g(s^2)$ enabling such $\Phi(s)$ should be infinitely continuous differentiable and to decrease monotonically from 1 to 0 while s^2 varies from 0 to $+\infty$. Two choices of g are given by Eqs. (5.18) and (5.19), respectively.

$$g(s^2) = \frac{1}{1 + s^2\lambda^2} \tag{5.18}$$

$$g(s^2) = e^{-s^2\lambda^2} \tag{5.19}$$

In case of Eq. (5.19) $\Phi(s)$ has a maximum at $|s| = \lambda$, with $\Phi'(s) > 0$ for $|s| < \lambda$ and $\Phi'(s) < 0$ for $|s| < \lambda$.

In the two-dimensional case, let ξ and η denote the local coordinates in directions perpendicular and parallel to ∇u respectively. Then the Perona-Malik equation can be rewritten as Eq. (5.20).

$$\partial_t(u) = g(|\nabla u|^2)u_{\xi\xi} + \Phi'(|\nabla u|)u_{\eta\eta} \tag{5.20}$$

The coefficient of $u_{\xi\xi}$ is always positive, hence (5.17) acts as a smearing filter washing details along the contour lines of the function u. The coefficient of $u_{\eta\eta}$ may be both positive and negative, hence, in the perpendicular (gradient) direction, slow gradient values are smeared out, while large gradient values (like edges) are sharpened instead of being blurred.

Both conduction coefficients lead to stable edges during image evolution. Regarding the difference between the two equations, wide regions are privileged over the smaller ones in case of Eq. (5.18), whereas high-contrast edges are privileged over low-contrast ones in case of Eq. (5.19).

Results of the conventional anisotropic diffusion (Perona and Malik) upon a gray scale image aiming edge enhancement are presented next. A 2D network structure of 8 neighboring nodes is considered for diffusion conduction. The parameters to be chosen for the diffusion are the number of iterations N, integration constant ΔT which is set usually to maximum value and the gradient modulus threshold that controls the conduction denoted by λ. Examples of how the filtering is performed on solar surface irradiance images are presented next. Both the conduction coefficients from Eqs. (5.18) and (5.19) were applied on the same satellite image, whereas the number of iteration, the integration constant and the gradient threshold yield the same values. The resulted images are shown in Fig. 5.10a, b, respectively. The same image details corresponding to the diffusion Eqs. (5.18) and (5.19) are illustrated in Fig. 5.10c, d.

As underlined in Fig. 5.10, the advantages of the proposed diffusion techniques are that edges remained stable over a very long time. It was demonstrated that edge detection based on this process clearly outperforms the linear Canny edge detector, even without applying non-maxima suppression and hysteresis thresholding. This is due to the fact that diffusion and edge detection are integrated in one single process instead of being treated as two independent processes subsequently applied. On the other side, the disadvantage of the Perona and Malik filter is the computational complexity of the iterative filtering which leads to increased computational time in case of real time processing or in case large amount of image data needs to be processed.

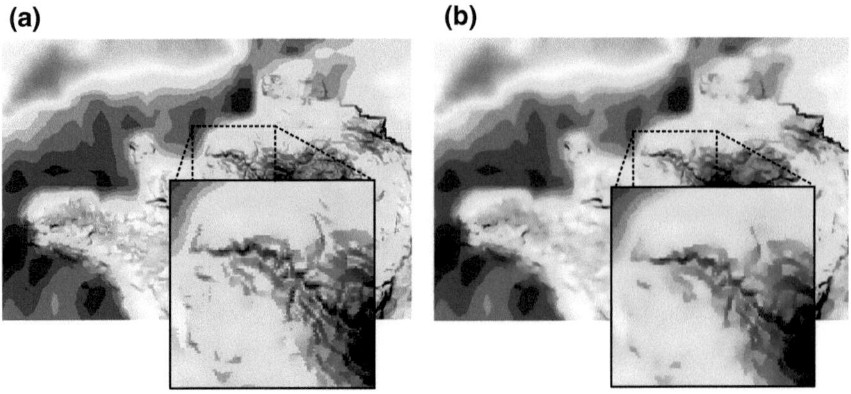

Fig. 5.10 Anisotropic diffusion applied for edge **a** Num_iter = 25, Kappa = 10, **b** Num_iter = 15, Kappa = 30, **c** Num_iter = 15, Kappa = 30, **d** Num_iter = 25, Kappa = 30

5.2.3 Hardware Implementation for Parallel Computation of Anisotropic Diffusion

Let $I = (p_{i,j})$ be the two dimensional array of intensities corresponding to the intensity levels within a grayscale satellite image I. Aiming noise removal and edge enhancement, anisotropic diffusion is applied and I_{diff} is obtained by evolving the original satellite image I, according to the diffusion process described by Eq. (5.16). Considering the discrete nature of the information within satellite image I, the aforementioned image is evolved using the discrete process described in the following section.

Let t = 1…N be the total number of iterations considered for the diffusion process. One of the iterations involves the following three steps:

- *Step 1*: for each pixel intensity $p_{i,j}$, using finite difference approximations, image gradients were computed on 8 different directions, relative to the cardinal directions (i.e. N, W, S, E) and intermediate cardinal directions (NW, NE, SW, SE). Thus, the gradients are denoted by η, whereas a lower index denotes the gradient directions (e.g. η_{NW} is the gradient on the NW direction computed by $\eta_{NW} = p_{i-1,j-1} - p_{i,j}$).
- *Step 2*: diffusion function c is evaluated for each direction, based on the computed gradient on the same direction:

$$c_N = e^{-(\eta_N/k)^2} \tag{5.21}$$

- *Step 3*: the pixel intensity $p_{i,j}^{t+1}$ within the resulted image I_{diff} after one complete iterations is computed as denoted by Eq. (5.13).

$$\begin{aligned} p_{i,j}^{t+1} &= p_{i,j}^{t} + c_N\eta_N + c_S\eta_S + c_E\eta_E + c_W\eta_W + \ldots \\ &\quad + 0.5c_{NE}\eta_{NE} + 0.5c_{SE}\eta_{SE} + 0.5c_{NW}\eta_{NW} + 0.5c_{SW}\eta_{SW} \end{aligned} \quad (5.22)$$

To perform the aforementioned computational steps, a custom processing hardware architecture was developed in order to achieve high-throughput (see Fig. 5.11). *Steps 1, 2* and *3* are parallelized for efficient computation using field programmable gate arrays (FPGAs) which enable spatial and temporal parallelism to be applied using the application specific hardware architecture development.

The functionality of the architecture for anisotropic diffusions is described. One of the iterations of the diffusion process is composed of five independent processing stages. The time intervals Δt_1 to Δt_5 correspond to the computational stages and they are performed by the proposed architecture as detailed in the Fig. 5.11. The digital logic elements for the processing steps are separated by vertical dashed lines.

Let *M* and *N* be the horizontal and vertical image size. The first computational stage within iteration *t* performs a local caching operation; *2M + 3* pixel intensity $p_{i,j}^{t}$ values are stored sequentially in a shift register.

The second computational stage (Δt_2) calculates image gradients in different directions using finite difference approximation. The gradient values (e.g. η_{NW} gradient on the NW direction) are stored in an intermediate register *RegA* to be used for further computations. The size of *RegA* is 8, as the number of all the gradient values η. The η_N, η_{NW}... η_W are computed in parallel using multiple sign adders. (e.g. $\eta_N = p_{i,j}^{t} - p_{i-1,j}^{t}$)

The third stage computes for each gradient values η stored in *RegA* its corresponding diffusion coefficient *c* (e.g. c_{NW} corresponds to the η_{NW} gradient value—see Eq. (5.14)). The computation of the exponential function is performed using

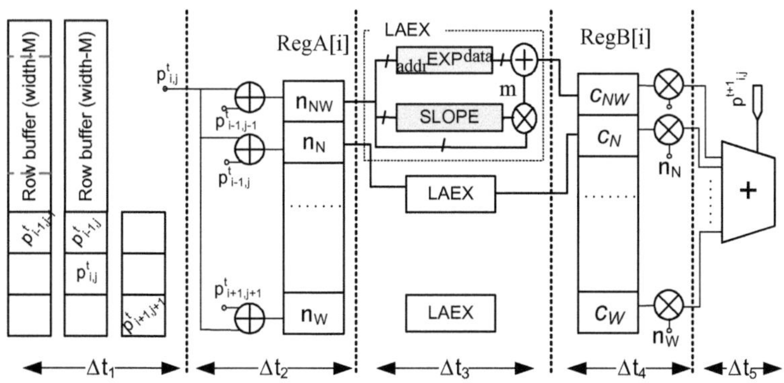

Fig. 5.11 Hardware Architecture for parallel Perona and Malik filtering

Table 5.3 Hardware resource usage for the anisotropic diffusion

No. of shift register cells (16 bits size)	No. of multipliers (16 bits)	No. of LUT based ROM (512 kB)	No. of adders (16 bits)
2.1 k	16	16	24

linear approximations. Such architecture is detailed in Chap. 3. The hardware implementation for diffusion function computation makes use of a look-up-table LUT-based ROM in order to store $A_i(x_i, y_i)$ and m_i values; A_i represents diffusion function values whereas m_i is the slope of two adjacent points A_i and A_{i+1}. These values are stored in EXP and SLOPE ROM memory. The processing unit for computing the diffusion coefficient is called LAEX (Linear Approximation Exponential) and delivers the exponential functions results according to Eq. (5.23).

$$f(x) = e^x = m_i(x - x_i) + y_i \tag{5.23}$$

The resulted diffusion coefficients c are stored in the *RegB* register and use further on to compute the output pixel intensity values $p_{i,j}^{t+1}$ which represents the pixel intensity values from the diffused image obtained at the end of iteration t.

The fourth stage computes the products between the register RegA[i] values and the computed diffusion coefficients c. A number of 8 multipliers are used in parallel for the aforementioned computation. The results simultaneously delivered by the multipliers are summed as described by the Eq. 5.15.

The fifth stage delivers the pixel intensity values $p_{i,j}^{t+1}$ of the output image at the end of the iteration t of the diffusion process.

Timing considerations for the proposed architecture are discussed next. Each pixel $p(i,j)$ from the image to be filtered is delivered each $3xT_{clk}$ cycles to the anisotropic diffusion architecture. The time intervals Δt_1 to Δt_5, equal to $3xT_{clk}$ each, allows the computational stages to be fully pipelined. Consequently, after an initial delay the pixel intensity values from the output image are delivered sequentially each $3xT_{clk}$ cycles. Table 5.3 illustrates the hardware resource usage for the proposed architecture for anisotropic diffusion implementation, considering a 1024×1024 gray-scale image.

5.2.4 Application-Specific Hardware Architecture for Perona and Malik Filter in Satellite Imagery—Case Study

The increase of the number of high resolution satellites into orbit and the number of applications which use satellite images leads to "big data" to be processed, which cannot be accommodated to the satellite's local computing infrastructures. The present section focuses on the importance of the proposed high-throughput computing architectures for processing satellite images. Consequently, the hardware architecture for the iterative Perona and Malik filter is presented in the context of the existing grid computing facilities, with regards to the computational time. In what follows, a description of grid-based solutions for satellite image processing is detailed.

The Large Hadron Collider (LHC) Grid is an emerging infrastructure developed by CERN that provides access to computing power and data storage distributed over the globe [http://wlcg-public.web.cern.ch/]. The grid infrastructure is organized according to a four tiered [20, 21]. A primary backup will be recorded on tape at CERN, the "Tier-0" centre of LCG (Grid). After initial processing, this data will be distributed to a series of Tier-1 centers, large computer centers with sufficient storage capacity for a large fraction of the data, and with round-the-clock support for the Grid. The Tier-1 centers will make data available to Tier-2 centers, each consisting of one or several collaborating computing facilities, which can store sufficient data and provide adequate computing power for specific analysis tasks. Individual scientists will access these facilities through Tier-3 computing resources, which can consist of local clusters in a University Department or even individual PCs, and which may be allocated to LCG on a regular basis.

In what follows it is described how the LCG Grid operates. The process begins with an individual user accessing a user interface (UI) through a personal account, with a user security certificate installed. The user describes a job that will run on the Grid. The job arrives at the Resource Broker (RB). A set of services running on the RB machine contribute to match job requirements to the available resources, schedule the job for execution to an appropriate Computing Element (CE). Each output of the user job performed by the CE is stored on a Grid Storage Element (SE). The computing elements are based on Intel Xeon Processors E5 families [22]. They are composed of up to 12 execution cores, each one supporting two threads. Regarding memory interfaces, the integrated memory controller supports 4 different 64 bits memory channels, whereas the processor frequency is up to 3 GHz. Following this approach, satellite images could be uploaded to the LHC Grid where they could be stored and processed, freeing up the satellite local computing resources. Additional processing strategies can be considered in case of satellite images. Thus, supplementary computational power can be added through application-specific architectures, using existing technologies such as FPGA or GPU (Graphic Processing Units). Both approaches make use of spatial and temporal parallelism in order to improve the computational performances for image processing algorithms.

Further on, the anisotropic diffusion algorithm is used for the exemplification of the speed-up factor introduced by the proposed application-specific hardware architecture, as compared to a general purpose processor (i.e. Xeon-based computing element). A detailed description of the algorithm is presented in Sect. 5.2.3. We consider one computational step to be composed of a memory read operation m_R, arithmetic operation o_{ALU} performed by arithmetic logic unit ALU or a memory write mW operation. Let C be the total number of computational steps necessary for one of the iterations of the anisotropic diffusion algorithm. Let $f_x = 3\ GHz$ and $f_a = 0.3\ GHz$ be the clock frequencies of the Xeon-based computing element and of the Virtex 5 FPGA, respectively. Moreover, let $S = M \times N\, pixels$ be the size of the gray-scale image onto which the number of C computational steps are applied. Taking into account that multiple architectures as the one proposed in this chapter can be instantiated, we perform the speed-up factor computation introduced by our approach as opposed to one execution core of the Xeon-based computing element. $P \times T^{x}_{clk}$ represents the number of clock periods necessary for the C computational steps performed by the Xeon-based computing element, whereas $Q\, x\, T^{a}_{clk}$ represents the number of clock periods necessary for the C computational steps performed by the proposed application-specific hardware architectures. Consequently, the speed up factor is given by:

$$F = P \cdot T^{x}_{clk} / Q \cdot T^{a}_{clk} \tag{5.24}$$

Considering the proposed architecture, after initial delay of $2M + 15$ clock cycles, pixel intensity values from the resulted image are delivered consequently each $3xT^{a}_{clk}$ clock cycles which leads to a total computational time of $Q = 3M \times N + 2M + 15$. On the other hand, the Xeon-based computing element performs the C number of computational steps as follows: 9 mR operations for the current and neighboring pixels used for the computation, 9 o_{ALU} for the finite difference approximations, 9 o_{ALU} operations for exponential function computation, 9 mW operations together with 9 mR operations for partial computation results storage, 9 o_{ALU} operations for multiplication, 9 o_{ALU} for the summation of the 9 terms for computing the resulted pixel intensity value and 1 mW operation.

Taking into account the timing considerations of the proposed architecture and the aforementioned timing considerations referred to the Xeon-based computing element, the resulted speed up factor is given by:

$$F = (64MxN) \cdot f^{a}_{clk} / (3MxN + 2M + 15) \cdot f^{x}_{clk} \tag{5.25}$$

Equation (5.25) leads to a speed-up factor F $\approx$ 2.13 for our proposed architecture as compared to a Xeon-based computing element.

5.3 Conclusions

Two PDE based image processing techniques are described and applied on microarray images and satellite images, respectively. Thus shock filter are successfully applied for microarray grid alignment and anisotropic diffusion is used for image preprocessing for feature enhancement. Both types of images are of large size, and, moreover, the algorithms for implementing the aforementioned image processing techniques are iterative, which lead to increased computational time. Consequently, hardware architectures are proposed for both shock filter and anisotropic diffusion implementation. The computational time for the proposed architectures are estimated and compared with state of the art approaches for parallel computation of image processing tasks. The results show the benefits of FPGA technology used for efficient implementations of image processing algorithms in terms of processing speed.

References

1. S. Osher, L.I. Rudin. Feature-oriented image enhancement using shock filters. SIAM J. Numer. Anal. **27**(4): 919{940, August 1990. 15, 43, 44, 45, 47, 48, 56, 57, 66, 129}
2. P. Perona, J. Malik, Scale-space and edge detection using anisotropic diffusion. IEEE Trans. Pattern Anal. Mach. Intell. **12**(7): 629{639, July 1990. 12, 15, 16, 17, 18}
3. B. Belean et al., Unsupervised microararray image segmentation for spot with irregular contours and inner holes. BMC Bioinformatics. **16**(412), 2015
4. J. Angulo, J. Serra, Automatic analysis of DNA microarray images using mathematical morphology. Oxford Bioinformatics **19**(5), 553–562 (2003)
5. Y. Wang, Q.M. Marc, K. Zhang, Y.F. Shih, A hierarchical refinement algorithm for fully automatic gridding in spotted DNA microarray image processing. Inf. Sci. Int. J. **177**(4), 1123–1135 (2007)
6. J. Angulo, J. Serra, Automatic analysis of DNA microarray images using mathematical morphology. Oxford Bioinformatics **19**(5), 553–562 (2003)
7. E. Zacharia, D. Maroulis, An Original genetic approach to the fully automatic gridding of microarray images. IEEE Trans. Med. Imaging **27**(6), 805–813 (2008)
8. L. Rueda, I. Rezaeian, A fully automatic gridding method for cDNA microarray images. BMC Bioinformatics **12**(113), 1–17 (2011)
9. D. Bozinov, J. Rahnenfuhrer, Unsupervised technique for robust target separation and analysis of DNA microarray spots through adaptive pixel clustering. Bioinformatics **18**(5), 747–756 (2002)
10. J. Angulo, J. Serra, Automatic analysis of DNA microarray images using mathematical morphology. Bioinformatics **19**(5), 553–562 (2002)
11. K. Zhang, H. Song, L. Zhang, Active contours driven by local image fitting energy. Pattern Recogn. **43**, 1199–1206 (2010)
12. R. Malutan, P. Gómez, M. Borda, Independent component analysis algorithms for microarray data analysis. Intell. Data Anal. **14**(2), 193–206 (2010)
13. B. Belean, M. Borda, B. Le Gal, R. Terebes, FPGA based system for automatic cDNA microarray image processing. Comput. Med. Imaging Graph. **36**(5), 419–429 (2012)
14. J.L. Bosch, F.J. Batlles, L.F. Zarzalejo, G. López, Solar resources estimation combining digital terrain models and satellite images techniques. Renew. Energy **35**, pp. 2853–2861 (2010)
15. J. Verbesselt, A. Zeileis, M. Herold, Near real-time disturbance detection using satellite image time series. Remote Sens. Environ. **123**, 98–108 (2012)
16. S. Fang, L. Xu, H. Pei, Y. Liu, Z. Liu, Y. Zhu, J. Yan, H. Zhang, An integrated approach to snowmelt flood forecasting in water resource management, industrial informatics. IEEE Trans. **10**(1), 548–558 (2014)
17. P. Guidotti, Y. Kim, J. Lambers, Image restoration with a new class of forward-backward-forward diffusion equations of Perona-Malik type with Applications to Satellite Image Enhancement. to appear in SIAM J. Imaging Sci. **6**(3), 1416–1444 (2013)
18. S. Bettahar, A. Boudghene Stambouli, An efficient PDE framework for satellite image classification, Revue des Energies Renouvelables. **13**(3), 369–377 (2010)
19. G. Huang et al., A LUT-based approach to estimate surface solar irradiance by combining MODIS and MTSAT data. J. Geophisical Res. 116:D22 (2011)
20. P. Buncic et al., LHC Computing Grid Project Report, Cern LCG, Geneva, CERN-LCG-2003-033 (2003)

21. J. Flix Molina, A. Forti, M. Girone, A. Sciab, Operating the worldwide LHC computing grid: current and future challenges 320th international conference on computing in high energy and nuclear physics. J. Phy. Conf. Ser. **513** (2014). doi:10.1088/1742-6596/513/6/062044
22. Intel, Case Study Intel Xeon Processor E5 Family High-Performance Computing, A new model for high-performance computing (2015)

Chapter 6
Efficient Hough Transform Implementation Using CAM Memories Applied on Satellite Imagery

Modern human civilization has a significant influence on the Earth, which leads to a strong need to assess the status of the natural environment. Earth Observation (EO) through remote sensing satellites gathers reliable information about the environment and provides the opportunity to minimize the negative impact of society and to improve social and economic well-being. Nevertheless, the evolution of satellite instrumentation leads to considerable satellite data volumes which require algorithms based on mathematics and statistics algorithms to extract valuable information, increased computational power for processing and also high-throughput and secured satellite data transmission for communication. EO data is commonly represented as satellite images. Image processing techniques based on partially differential equations (PDE), artificial intelligence (AI) and feature extraction techniques (e.g. Hough transform) are available for permanent monitoring of the Earth's land and oceans. The variety, complexity and the iterative nature of such image processing algorithms demand also for efficient implementations in terms of computational power. Consequently, the present chapter proposes a FPGA-based approach for efficient computation of image processing algorithms applied for satellite image analysis.

The image processing task discussed within this chapter is image segmentation, using circular Hough transform. The main idea is to extract circular shapes from the image using the Hough transform. It is well known the increased computational power needed by the Hough transform, due to recurrent search in the Hough space of various shapes (e.g. in our case circular objects). In our case, a content addressable memory (CAM) is proposed for efficient circular feature extraction from satellite images. Further on, an introduction to satellite imagery for oil slick detection is presented, followed by the application of Hough transform to satellite image analysis (detection of seepage oil slicks). The computational time for the circular feature extraction is estimated and, an approach for efficient feature extraction based on CAM memories is proposed.

B. Belean, *Application-Specific Hardware Architecture Design with VHDL*, Signals and Communication Technology,
https://doi.org/10.1007/978-3-319-65025-8_6

6.1 Satellite Imagery for Oil Slick Detection

Satellite images are known to be very effective at observing oil slick both generated by oil reservoirs beneath the oceans and by oil spills due to pollution. In case of beneath the ocean oil reservoirs, the surface expression of how petroleum migrates from beneath the ocean towards its surface is known as oil seepage. This is helpful in localizing any sort of oil accumulation beneath the bottom of the oceans. In other words, the oil is transported towards the ocean surface by oil-coated gas bubbles. At the surface of the ocean, a thin oil film is formed, whereas gas bubbles are lost in the atmosphere. The oil film (seep) represents detectable volumes of oil and gas, which in calm sea conditions are described by concentric shapes that change color depending on the angle of view. The pollution slicks are thicker than the seepage slicks, both of them being interpreted as dark patches on the satellite imagery.

Considering image processing algorithms, a topic of current research is to monitor oil pollution and the sea-surface expression of beneath the ocean oil accumulation through oil slick detection/segmentation applied on synthetic aperture radar (SAR) images [1, 2]. Environment monitoring via radar imaging may benefit from state-of-the-art image processing algorithms based on PDEs for denoising, fusion and segmentation of satellite data. Sophisticated approaches like neural networks together with computationally expensive Hough transforms are applied for segmentation and fault tracking in satellite image data volumes [3–5]. When satellite image time series are envisaged to extract reliable information, grouped frequent sequential patterns (SP) are successfully applied for segmentation and feature extraction [6] and showing great potential in SAR image time series segmentation for oil slicks detection. Considering the complexity of such methods and the large volumes of available satellite data, parallel and distributed computing approaches are needed to tackle the Big Data paradigm of satellite data processing [7]. Consequently, parallel computing approaches are accounted for development of application specific-hardware architectures for satellite imagery.

We have seen so far that complex image processing algorithms were successfully used for the detection of oil slick generated by both pollution and seepages. Before proposing a solution to improve the computation efficiency of such computationally expensive image processing tasks, we first explain in detail the principle of Hough transform and how it is applied for circular feature extraction. In our case, we aim to detect the oil seepages, which in calm sea conditions are described by concentric circular shapes on the sea surface. The concentric shapes are visible as circular dark patches on the satellite images (see Fig. 6.1).

6.1.1 Circular Hough Transform

The circular dark patches on the satellite images are to be localized using the Hough transform. The basic principle is described next: the Hough transform is a procedure which detects the occurrences of a shape in an image. The a priori assumption is

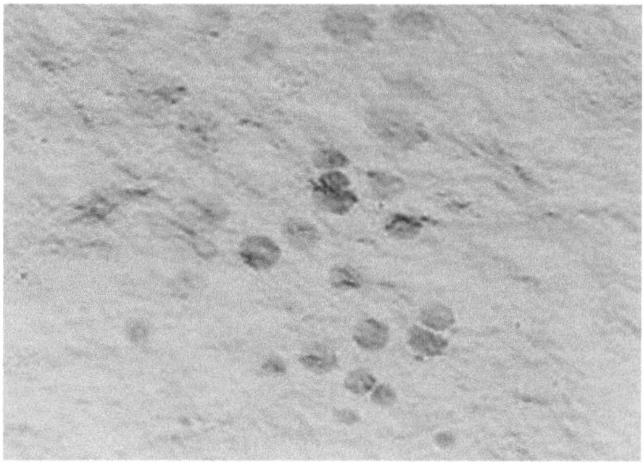

Fig. 6.1 Example of circular oil slick generated by oil seepages

that the shape can be described in a parametric form. Thus, in our case a 2D circle can be parameterized as denoted by Eq. (6.1).

$$(x-a)^2 + (x-b)^2 = r^2 \tag{6.1}$$

It is important to know that, the edges within the image to be processed constitute the a priori information for the Hough transform. Thus, an edge detector operator (e.g. Canny filter) is first use for the identification of image edges, and consequently the boundaries for the image edges. Once the edges are detected, we assume that an edge point is part of a circle. This leads to the following case: the point could belong to a unique family of circles with varying parameters (a, b, r). The (a, b) represent the centre of the circle whereas r corresponds to its radius. The procedure of the Hough transform follows next. An accumulator array whose axis are the (a, b, r) parameters is created. Each determined edge point (x_i, y_i) votes within the accumulator array (see Fig. 6.2). At the end of the voting procedure, the maximum values within the accumulator corresponds to the detected circular shapes having the (a, b, r) parameters.

As shown before, the basic principle of the Hough transform is not necessary a very complex one. Nevertheless, the multiple search for circular shapes with various parameters together with the voting procedure for each circular shape lead to an increased computational complexity, especially in the case of large size images. Aiming to overcome this disadvantage, further on we propose an approach for efficient computation of the Hough transform using FPGA technology.

6.1.2 CAM-Based Approach for Efficient Hough Transform Implementation

Further on, we focus on the adaptive Hough transform implementation for slick detection and segmentation, which benefits of the parallel computation capabilities

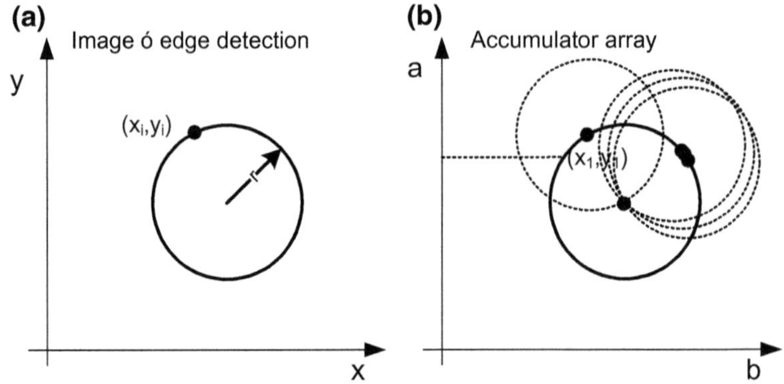

Fig. 6.2 The Hough transform applied for circular shape detection; **a** original image with detected edge, **b** the accumulator array for the voting procedure

of the FPGA technology. The main scope is to reduce the computational complexity of the circular Hough transform. The novelty within the proposed segmentation approach is represented by the content addressable memory (CAM) used for efficient voting in the Hough space. The approach is used for seepage detection. Considering the dependency of oil slicks generated by beneath oceans reservoirs on local wind conditions, a permanent search on satellite SAR images is envisaged in order to capture the circular gas bubbles accumulation, expressed as circular oil slicks. Thus the computational cost of image processing methods such as partial differential equations and Hough is high; consequently a strong motivation for developing hardware implementation of such algorithms exists.

Thus, for circular oil slicks detection the Hough transform is considered prior to an edge detection step. The CAM based approach for circular shape extraction which performs efficient computation of the Hough transform by enhanced memory access is presented in what follows.

Considering the SAR image space $I_S = p(x,y)$ the two dimensional array of pixel intensity values, we define a circle as $(x-a)^2 + (y-b)^2 = r^2$ to describe circular shapes of oil slicks. The Hough transform implementation through CAM memories is detailed in case of the circular shape detection. Firstly, it is to be mentioned that CAM memories are known for their special features of returning the address for a given value found in the memory and delivered at CAM input. Thus, the main benefit is that no memory search is needed for returning the address of a specific value from the memory. In our case, multiple CAM memories are used as processing elements (PE) which fulfill the following tasks: accumulator for computing the Hough space through voting and a decision circuit which specifies if voting is performed or not. For a better understanding, we present in more detail the approach for parallel computing of Hough transform using CAM memories.

A gradient based approach is used to select edge points (x_i, y_i) within the image space. The circle equation is written as the following linear recursive Eq. 6.2.

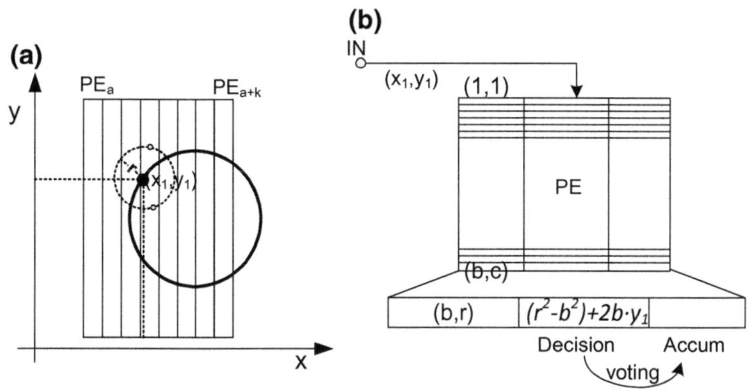

Fig. 6.3 Principle of the hardware implementation for the Hough transform based on CAM: *left* processing elements PE are built for image lines, *center* CAM based element performs parallel voting in the Hough Space which is stored in accumulators, *right* examples of curved oil slicks (morphological thinning is to be applied for detection)

$$(x - a)^2 + y^2 = (r^2 - b^2) + 2by \tag{6.2}$$

Let the Hough space be the total number of circles determined by the *(a,b,r)* parameter space for all the edge points $(x_i.y_i)$. A PE_a is taken into account for each image column in the Hough space. Each *PE* is addressed by the parameter *a* as denoted by Fig. 6.3. Each edge point (x_i,y_j) is delivered as input to the decision circuit of the PE_a, which computes in parallel the corresponding $(r^2-b^2) + 2by_1$ values for each pair *(b,r)*. Parallel search is computed for the $(x_i-a)^2 + y_j^2$ values for all the decision circuits, and the accumulators are incremented where the circle equation is verified. In this way the Hough space is computed and the circles are detected in the final computing step by detecting for each PE the CAM words in the accumulator for which the accumulator values are above a threshold value v_{thr}.

6.2 Memory Implementation Using FPGA

Memories represent mandatory components in any digital signal processing system used for local data storage. By local data storage one can understand buffers, stacks, shift register, delay lines, waveform storage and generation, function tables or program storage for embedded processors needed by the digital system in question to function properly. The term memory actually is the shorthand for *physical memory*, meaning any chip capable of holding data. There are various types of memories, depending on their key design metrics such as memory density, access time, power dissipation. Further on, a short memory classification is presented as a general topic, whereas the available memories within an FPGA chip (i.e. distributed memory and block RAM based memories) are detailed. Note that block RAMs are intrinsic components of the FPGA chip, and they represent random access memory blocks placed at the edge sides of the chip (see Chap. 1).

6.2.1 Memory Types

Considering the possibility to store data even after power removal, memories are classified as *volatile* or *non-volatile*. The non-volatile ones retain their stored information after the removal of power. The most common volatile and non-volatile memories are random access memories (RAM) and read-only memories (ROM), respectively. RAM memories are further on classified as static or dynamic. The dynamic ones need periodic refresh but they are simpler and with higher density. Their common characteristic is that memory locations can be read or written in a random order. On the other hand, the ROM memories are another form of data storage that cannot be easily altered or reprogrammed. There are three types of ROM memories: programmable read only memories (PROM), erasable programmable read-only memory (EPROM) and electrically erasable programmable read-only memory (EEPROM).

When considering FPGA-based memories employed in digital designs, there are two types of FPGA components to be used in order to build memories: look-up tables (LUT) and block RAMs (BRAM).

Distributed RAM memories can be built using LUTs. Two LUTs from an FPGA configurable logic block (CLB) can make a 32×1 single port distributed RAM. The memory type is called *distributed* due to the fact that CLB, and implicit LUTs that are used for memory construction are spread all over the FPGA chip. The distributed RAM size can be increased by cascading multiple LUTs [8]. As far as for the read/write operations, the distributed RAM read is asynchronous, whereas the write operation is synchronous. A synchronous read can be done using extra flip-flops.

Block RAM memories are dedicated blocks of memories within the FPGA chip. They represent the efficient memory implementation for the most memory requirements. Synchronous write and read operations are performed.

6.2.2 Inferred and Instantiated Memories Using VHDL

Considering the VHDL code to be used for building memory blocks, there are two methods for handling this: instantiation and inference.

The VHDL code synthesizer automatically configures the LUTs as function generators or ROM memories when needed. Moreover, if small array of registers are used within the design, the synthesizer configures LUTs as distributed RAM memories for the registers implementation. In case the registers are of increased size, the synthesizer normally uses block RAMs instead of LUTs to implement the registers functionality. This synthesizer automatic behavior is called inference. In case the designer is not satisfied with the synthesizer way of inferring the logic blocks, he has the option to override the logic blocks behavior by using constraints. Another option is to instantiate library parts to force the digital logic to be created as specified in library definition. This procedure of using library parts is called *instantiation* [9]. Further on VHDL code examples are detailed for both inferred and instantiated RAM memories.

Inferred distributed RAM

A 256 words distributed RAM memory is inferred by the following code description. The *we* port indicates a read or write operation depending on the logic value *0* or *1*, respectively. The *data_in* indicates the word to be written in the RAM at the *addr* address, whereas the *data_out* indicates the word read from the *addr* address. Next the entity declaration is listed.

```
---------- Entity declaration--------------------------------
LIBRARY ieee;
USE ieee.std_logic_1164.all;
USE ieee.std_logic_arith.all;
USE ieee.std_logic_unsigned.all;
entity inferdRAM is
     generic (bits : integer : = 32;
     addr_bits: integer: = 8);
     port (  clk:    in std_logic;
             we:    in std_logic;
             addr:  in std_logic_vector(addr_bits-1 downto 0);
             data_in: in std_logic_vector(bits-1 downto 0);
             data_out:   out std_logic_vector(bits-1 downto 0));
end inferRAM;
---------- End of entity declaration--------------------------
```

The behavioral description of the distributed RAM with asynchronous read is listed next:

```
---------- Behavioral declaration of RAM----------------------
architecture behavioral of inferdRAM is
type ram_type is array (2**addr_bits-1 downto 0) of std_logic_vec-
tor (bits-1 downto 0);
     signal RAM : ram_type;
begin
     process (clk)
         begin
            if (clk'event and clk = '1') then
```

```
            if (we = '1') then
               RAM(conv_integer(unsigned(addr))) <= data_in;
            end if;
         end if;
      end process;
   data_out <= RAM(conv_integer(unsigned(addr)));
end behavioral;
---------- End of behavioral declaration---------------------------------
```

Note the *data_out* assignment is performed outside the *clk* process, meaning an asynchronous read is performed. Placing the *data_out* assignment inside the *if (clk'event and clk = '1')* statement leads to a distributed RAM memory with fake synchronous read. The inferred distributed RAM with synchronous read looks like depicted in Fig. 6.4.

Inferred block RAM

In order to have a block RAM inferred to a VHDL memory description, the read address must be registered on the RAM clock edge. The VHDL code example s given next:

```
---------- Entity declaration-------------------------------------------
LIBRARY ieee;
USE ieee.std_logic_1164.all;
USE ieee.std_logic_arith.all;
USE ieee.std_logic_unsigned.all;
entity inferBRAM is
   generic (bits : integer : = 32;
   addr_bits: integer: = 8);
   port ( clk:  in std_logic;
         we:   in std_logic;
          addr:  in std_logic_vector(addr_bits-1 downto 0);
            data_in: in std_logic_vector(bits-1 downto 0);
          data_out:  out std_logic_vector(bits-1 downto 0));
end inferRAM;
---------- End of entity declaration------------------------------------
```

The presented entity declaration corresponds to a 256 × 32 bits word block RAM memory. The *we* port indicates a read or write operation depending on the

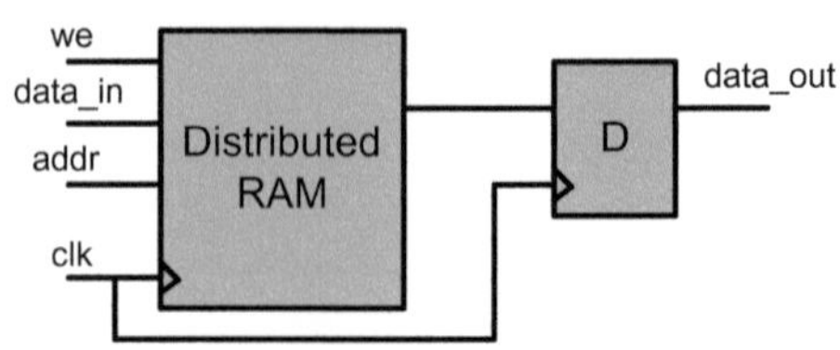

Fig. 6.4 Distributed RAM with synchronous read

logic value *0* or *1*, respectively. The *data_in* indicates the word to be written in the RAM at the *addr* address, whereas the *data_out* indicates the word read from the *addr* address. The following behavioral description assigns the read address on the memory clock edge through the *read_addr* signal.

```
——— Behavioral description of Block RAM———
architecture behavioral of inferdRAM is
type ram_type is array (2**addr_bits-1 downto 0) of std_logic_vec-
tor (bits-1 downto 0);
      signal RAM : ram_type;
      signal read_addr: std_logic_vector(addr_bits-1 downto 0);
begin
      process (clk)
          begin
             if (clk'event and clk = '1') then
                if (we = '1') then
                   RAM(conv_integer(unsigned(addr))) <= data_in;
                end if;
                read_addr <= addr;
             end if;
          end process;
      data_out <= RAM(conv_integer(unsigned(read_addr)));
end behavioral;
——— End of behavioral declaration———
```

Inferred ROM

An example of 8 × 16 ROM memory is given next. Note that memory initialization is done trough constant declaration. The read operation is performed in a concurrent manner (i.e. no clock signal involved) by the following code line: *data_out <= ROM(conv_integer(unsigned(addr)));*.

```
——— Entity declaration———
LIBRARY ieee;
USE ieee.std_logic_1164.all;
USE ieee.std_logic_arith.all;
USE ieee.std_logic_unsigned.all;
entity inferROM is
      generic (bits : integer : = 16
      addr_bits: integer : = 3);
      port ( addr:    in std_logic_vector(addr_bits-1 downto 0);
              data_out: out std_logic_vector(bits-1 downto 0));
end inferROM;
architecture behavioral of inferROM is
```

```
type rom_type is array (2**addr_bits-1 downto 0) of std_logic_vec-
tor (bits-1 downto 0);
     constant ROM: rom_type:=
              (X"0000",
               X"1234",
              X"5678",
              X"9ABC",
              X"DEF0",
              X"EF00",
              X"F000",
              X"0000");
begin
     data_out <= ROM(conv_integer(unsigned(addr)));
end behavioral;
---------- End of behavioral declaration--------------------------
```

Instantiated block RAM

Further on an example for the instantiation of the RAMB16_S9 block RAM memory is provided to underline how dedicated random access memory blocks with synchronous write capability can be used [10]. The block RAM memory configuration includes 2048 × 8 data cells, 2048 × 1 parity bits, 11 bits address bus, 8 bits data bus and 1 bit parity bus. An overview of the component can be seen in Fig. 6.5.

The functionality of the RAMB16_S9 is summarized as follows. The *Low* value on the *EN* port means no data is written and the outputs (*DO* and *DOP*) retain the last state. When *EN* is *High* and reset (*SSR*) is *High*, *DO* and *DOP* are set to *RVAL*. When *EN* is *High* and *WE* is *Low*, the data stored in the RAM address *ADDR* is read during the rising clock edge. When *EN* and *WE* are *High*, the data on the data inputs (*DI* and *DIP*) is loaded into the word selected by the write address (*ADDR*) during the rising clock edge and the data outputs (*DO* and *DOP*) reflect the addressed word.

The VHDL code for RAMB16_S9 memory instantiated is given next, whereas the component attributes are also underlined.

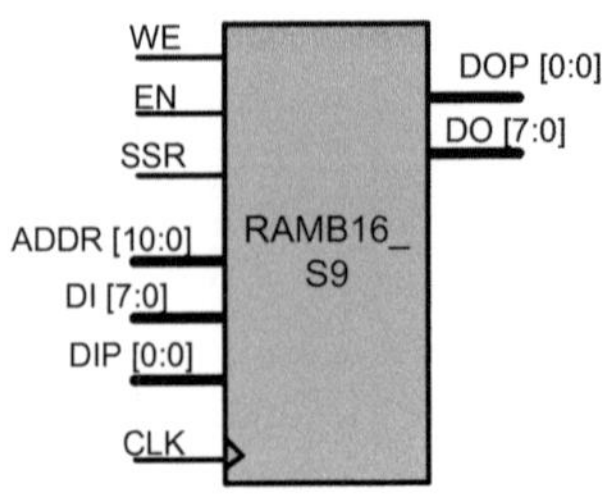

Fig. 6.5 Block RAM component RAMB16_S9 configuration (Spartan 3 FPGA)

```
——— instantiation of an RAMB16_S9 BRAM memory——————————
RAMB16_S9_inst: RAMB16_S9
generic map (
    INIT => X"000",
    SRVAL => X"000",
    WRITE_MODE => "WRITE_FIRST",
    INIT_00 =>
X"0000000000000000000000000000000000000000000000000000000000000000",
    INIT_01 =>

X"0000000000000000000000000000000000000000000000000000000000000000",
    ...
    INIT_3F =>

X"0000000000000000000000000000000000000000000000000000000000000000",
    INITP_00 =>

X"0000000000000000000000000000000000000000000000000000000000000000",
    INITP_07 =>

X"0000000000000000000000000000000000000000000000000000000000000000")
    port map (
      DO => DO, -- 8-bit Data Output
      DOP => DOP, -- 1-bit parity Output
      ADDR => ADDR, -- 11-bit Address Input
      CLK => CLK, -- Clock
      DI => DI, -- 8-bit Data Input
      DIP => DIP, -- 1-bit parity Input
      EN => EN, -- RAM Enable Input
      SSR => SSR, -- Synchronous Set/Reset Input
      WE => WE -- Write Enable Input);
———— end of the RAMB16_S9 memory instantiation————————
```

In order to specify the parameters of the instantiated memory, the following attributes are availabele for the instantiation: *INIT*, *INIT_00* to *INIT_3F*, *INITP_00* to *INITP_07*, *SRVAL* and *WRITE_MODE*. The *INIT* attributes specify the initial contents of the RAM memory, *SRVAL* allows the data output of the memory *DO* to be initialized with either *‚0'* or *‚1'* logic value after a reset (*SSR*), whereas the WRITE_MODE attribute specifies the behavior of the *DO* port upon a write command (the common behaviour is to keep the previous value on the output port and wont update the output port upon a write command).

6.2.3 Memory Organization

We've seen so far that, internal FPGA memories can be built using LUT of the configurable logic blocks or the specific memory blocks namely block RAMs. Depending on the VHDL code, LUT are used for inferring either ROM, distributed RAM or local register, whereas block RAMs may also used for inferring RAM memory. Moreover, instantiation of different memory components can be performed based on the library components used by the VHDL synthesizer (e.g. block RAM instantiation). Each of these memories can be used either as shift registers, circular registers, first-in first-out (FIFO) memories or stacks. For example, we can add specific functionality to a register such that data is written into the first register cell and consequently the other cell content is shifted to the right whereas the last cell content is delivered as output data. In this manner the initial register becomes a shift register. Adding specific functionalities to the memory blocks belongs to the concept of *memory organization*. Classic approach to organize memory is to build components such as shift registers, FIFO memories or even content addressable memories. The description and code example for shift registers and FIFO memories are given within current section. For the CAM memory, employed for Hough transform implementation, a separate section is dedicated (Sect. 6.3). Within this section, the main principle of CAM memories together with an example of VHDL code for CAM memory implementation is provided.

Shift registers are memory units which store n words with synchronous access for read and write operations. For each word input, another word is delivered as output data. The logic block for a shift register can be depicted in Fig. 6.6.

The VHDL code for the *n* words shift register inference starts with the *shift_register* entity, which includes a clock signal port *clk*, a *data_in* input port for the 8 bits word and a *data_out* output port to return the output word. The architecture description for the *shift_register* entity is also detailed, where the input output assignments are performed and also, the words are shifted within the register according to the *for loop* statement. Note that all the assignment are performed on clock signal edge (i.e. synchronous behavior).

```
———————— VHDL code for the shift register————————————————————————
library ieee;
use ieee.std_logic_1164.all;
entity shift_register is
     generic( n : integer : = 16 );
```

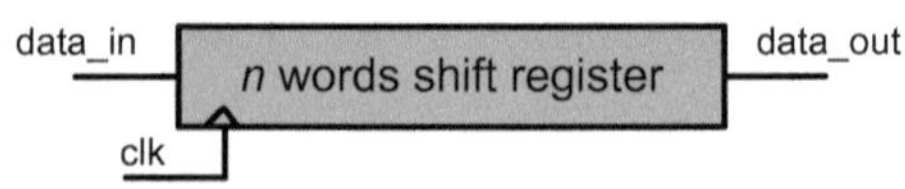

Fig. 6.6 Shift register

```
     port     ( data_in  : in std_logic_vector (7 downto 0);
              data_out : out std_logic_vector (7 downto 0);
                  clk  : std_logic);
end entity shift_register;

architecture reg of shift_register is
      type data_array is array
( 0 to n-1 ) of std_logic_vector (7 downto 0);
      signal mem: data_array;
begin
      process( clk )
      begin
          if (clk'event and clk = '1') then
             mem(0)  <= data_in;
             data_out  <= mem(n-1);
             for j in 1 to n-1 loop
                 mem(j)  <= mem(j-1);
             end loop;
          end if;
      end process;
end reg;
---------- end of the shift register description----------------------------
```

FIFO memories are memory units which store *n* words, having separated read and write ports. Both synchronous and asynchronous read and write operation may be performed. Their specific behavior is given by the *full* and *empty* flags which specify if the memory is either full or empty. The VHDL code for the *n* words FIFO memory inference starts with the *FIFO_mem* entity, which includes a clock signal port *clk*, a *data_in* input port, a *data_out* output port, read and write input ports to specify the memory operation to be performed and also the two output ports for the empty and full flags.

```
---------- VHDL code for the FIFO memory----------------------------
library ieee;
use ieee.std_logic_1164.all;
entity FIFO_mem is
      generic( n : integer : = 16);
      port(  data_in : in std_logc_vector (7 downto 0);
             data_out: out std_logc_vector (7 downto 0);
             clk, read, write, reset : std_logic;
             full, empty : out std_logic);
end entity FIFO_mem;
```

```
architecture fifo of FIFO_mem is
    subtype index is natural range 0 to n-1;
    type data_array is array( 0 to n-1 ) of std_logc_vector (7 downto 0);
    signal mem: data_array;
    signal idx : index;
begin
    process(clk)
    begin
        if reset = '1' then
          idx <= 0;
        elsif clk'event and clk = '1' then
            if read = '1' then
                if (idx /= 0 ) then
                    data_out <= mem (idx);
                    empty <= '0;'
                    idx <= idx - 1;
                end if;
            end if;
            if write = '1' then
                if ( read_idx < n-1 ) then
                    data_out <= mem(idx);
                    empty <= '0;'
                    full <= '0';
                    idx <= idx + 1;
                end if;
            end if;
            if ( idx = 0 ) then
                empty <= '1;'
            end if;
            if ( idx = n-1 ) then
                full <= '1;'
            end if;
    end if;
  end process;
end fifo;
———— end of the FIFO memory description————————————
```

Concerning the VHDL code description for the FIFO memory, an asynchronous reset initializes the FIFO index *idx* with zero. The *idx* signal specifies the curent position from the FIFI where the read and write operation are performed. All the other assignments of the FIFO logic block are performed on clock event. Thus, in case the fifo is not empty, *read_idx* < *n−1*, a write operation can be done, if a *'1'* logic is present at the *write* input port; the index *idx* is incremented. A read operation is performed if a *'1'* logic is present at the *read* input port; the index *idx* is decreased. Moreover, the flags *empty* and *full* are assigned with *'1'* logic value, if the memory is empty or full. In all other cases the flags are assigned with *'0'*.

6.3 CAM Memory Implementation Using VHDL

In order to find a given word *w* within *n* words depth ordinary memory, O(n) computational steps are needed. For an efficient search of the word *w*, specific search procedures such as binary trees can be employed, reducing the order of growth for the computational complexity to O(log n). A more efficient approach is to use CAM memories, for which each memory location is searched in parallel. Thus a search key (i.e. one word of data) is delivered as input to the CAM memory which returns whether or not a match occurred and also the index of the matched word within the memory (i.e. the memory address of the match). This is performed within a single computational step. The main disadvantage is that such memory is expensive to implement in terms of hardware usage. A number of *n* comparators are needed for the implementation, *n* representing the memory depth.

The following VHDL code corresponds to an inferred CAM memory. The memory depth is of size *n*. The CAM word size is *8* bits length. The input ports *key*, *search* and *reset* are used as follows: the *'0'* or *'1'* values on the *search* input port correspond to write or match operation for the CAM memory. Thus, while the CAM is not *full* and if *search* is *'0'* then write operations are performed in the CAM. In case *search* is *'1'*, then the *key* input is compared with the whole CAM content, and if a match is found, the memory returns *'1'* on *found* output port.

```
―――――――― VHDL code for the CAM memory――――――――――――――――――――――
library ieee;
use ieee.std_logic_1164.all;
use ieee.std_logic_arith.all;
entity cam is
generic( n : positive : = 8 );
port( key : in std_logic_vector (7 downto 0);
      found : out std_logic;
      search : in std_logic;
      full : out std_logic;
      reset : in std_logic;
      clk : in std_logic);
end entity cam;
```

The behavioral description of the CAM memory is described in the next VHDL code section. The variable *match_array* is used to determine if a match between the key and CAM content exists on each CAM memory location. Consequently, the *match_array* contains *n* logic values which represent if a match is found at the memory location *i*, with *i* from *0* to *n*. Finally, an OR logic between all *match_array* values determines if the key is founded within the CAM memory content.

An asynchronous *reset* initializes the *full* flag with *0* and also the *write_addr* to 0. In case *search* input is *'0'* a write operation is performed at *write_addr* memory location and also the *write_addr* is incremented for the next write operation. In case *search* input is *'1'*, the *key* input is looked for in the CAM memory content. The *found* output is assigned with *'0'* or *'1'* whether a match is found or not.

```
architecture cam_memory a of cam is
     type data_array is array(0 to n-1) of std_logic_vector (7 downto 0);
     type bool_array is array( 0 to n-1 ) of boolean;
     signal mem: data_array;
     signal write_addr : natural;
begin
process( clk, reset )
variable match : boolean;
variable match_array: bool_array;
begin
     if reset = '1' then
       write_addr <= 0;
       full <= '0';
     elsif clk'event and clk = '1' then
            if search = '1' then -- search mode
              for j in data_array'range loop
                  match_array (j) : = (mem(j) = key);
              end loop;
              match : = match_array(0) or match_array(1) or
              match_array(2) or match_array(3) or match_array(4) or
              match_array(5) or match_array(6) or match_array(7);
              if match then found <= '1';
                  else found <= '0';
                  end if;
              else -- add a new entry
                  if write_addr = data_array'high + 1 then
                     full <= '1';
                  else
                     mem( write_addr ) <= key;
                     write_addr <= write_addr + 1;
                  end if;
              end if;
```

```
            end if;
        end process;
end cam_memory;
———————— end of VHDL code for the CAM memory————————————————————
```

Further on, a test-bench for testing the CAM memory functionality is provided together with the simulation results (Fig. 6.5). The *test_cam* is the entity name used for testing. Inside the *test_cam* architecture the cam component is defined. Moreover, *key, search, reset, clk, found* and *full* signals are defined in order to provide inputs to the cam entity under test and to read the outputs.

```
use ieee.std_logic_1164.all;
use ieee.numeric_std.all;
entity test_cam is
end test_cam;
architecture behavior of test_cam is
component cam
     port(
          key : in std_logic_vector(7 downto 0);
          found : out std_logic;
          search : in std_logic;
          full : out std_logic;
          reset : in std_logic;
          clk : in std_logic
         );
     end component;
signal key : std_logic_vector(7 downto 0) : = (others => '0');
signal search : std_logic : = '0';
signal reset : std_logic : = '0';
signal clk : std_logic : = '0';
signal found : std_logic;
signal full : std_logic;
constant clk_period : time : = 10 ns;
```

Within the test-bench architecture, the CAM memory is instantiated and also inputs are assigned; the outputs are monitored using the simulation results provided in Fig. 6.7. Thus, a *reset* is applied on the proposed CAM memory, followed by memory write operations until the CAM memory is full (the full state is signalized at the *full* output port). The proposed test-bench architectures writes 8 bits integer values from *1* to *8* in the CAM memory. Once we have a full CAM memory, various keys are looked up. The output *found* signalizes if a match of the input *key* (8 bits input) exists within the memory. In our case the memory content is initialized with 8 bits integer values from 1 to 8. Consequently, the memory returns a match was found in case the *key* delivered as input is either 4 or 1.

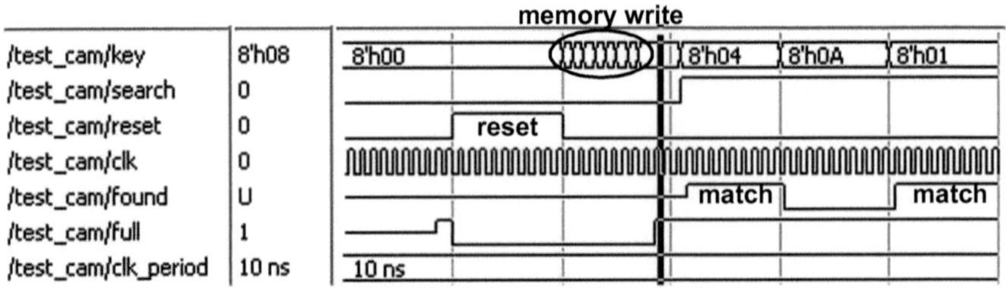

Fig. 6.7 Simulation results considering the proposed CAM memory implementation and the detailed test-bench architecture

```
begin
   uut: cam port map (
       key => key,
       found => found,
       search => search,
       full => full,
       reset => reset,
       clk => clk
           );
clk_process :process
variable depth: integer: = 8;
begin
   clk <= '0';
   wait for clk_period/2;
   clk <= '1';
   wait for clk_period/2;
   end process;
   stim_proc: process
   begin
   key <=x"00";
   reset <= '0';
   wait for 100 ns;
   reset <= '1';
   wait for 100 ns;
   reset <= '0';
   search <='0';
   for depth in 1 to 8 loop
   key <=std_logic_vector(to_unsigned(depth,8));
   wait for clk_period;
   end loop;
   wait for clk_period*3;
```

```
    search <='1';
    key <= x"04";
    wait;
end process;
end;
```

6.4 Conclusions

The main benefit of using CAM memories is that they offer efficient search possibility by delivering the address for a specific memory content delivered as input data. The main disadvantage is the increased hardware usage for this type of memories. Consequently, in case of search procedure within the image space, an increased size image leads to increased CAM memory depth for efficient search algorithms implementation. Nevertheless, spatial image partitioning strategies such as Voronoi diagrams may be employed in order to reduce the image size and also the CAM memory depth. The proposed approach makes use of the CAM memories to extract circular shape from images. Future work aims to extend CAM based search for various type of image features such as elliptical shape or parameterized curves.

References

1. H. Song, B. Huang, K. Zhang, A globally statistical active contour model for segmentation of oil slick in SAR imagery. IEEE J. Sel. Top. Appl. Earth Obs. Remote Sens. **6**(6), 2402–2409 (2013)
2. G. Suresh, C. Melsheimer, J.H. Körber, G. Bohrmann, Automatic estimation of oil seep locations in synthetic aperture radar images. IEEE Trans. Geosci. Remote Sens. **53**(8), 4218–4230 (2015)
3. Gui Gao, Lingju Zhaoa, Jun Zhang, Diefei Zhou, Jijun Huang, A segmentation algorithm for SAR images based on the anisotropic heat diffusion equation. Pattern Recogn. **41**(10), 3035–3043 (2008)
4. S. Singh, T.J. Bellerby, and O. Trieschmann, Satellite oil spill detection using artificial neural networks. IEEE J. Sel. Top. Appl. Earth Obs. Remote Sens. **6**(6), (2013)
5. B. Belean, C. Belean, C. Floare, C. Varodi, A. Bot, G. Adam, Grid based high performance computing in satellite imagery. Case study—Perona–Malik filter. Comput. Res. Model. **7**(3): 399–406 (2015)
6. A. Julea, N. Méger, Ph. Bolon, C. Rigotti, M.-P. Doin, C. Lasserre, E. Trouvé, and V. Lăzărescu, Unsupervised spatiotemporal mining of satellite image time series using grouped frequent sequential patterns. IEEE Trans. Geosci. Remote Sens. **49**(4): 1417–1430 (2011)
7. I. Zinno, L. Mossucca, S. Elefante, C. De Luca, V. Casola, O. Terzo, F. Casu, R. Lanari, Cloud computing for earth surface deformation analysis via spaceborne radar imaging: a case study. IEEE Trans. Cloud Comput. 4(1) (2016)

8. P.P. Chu, *Introduction to Digital System Design—Chapter 1 from RTL Hardware Design Using VHDL: Coding for Efficiency, Portability and Scalability* (Cleveland State University, Ohio, Wiley, 2006)
9. P. Eles, Z. Pang, System Synthesis with VHDL, Springer—Science + Business Media, March 2000, Presentation based on: System Synthesis with VHDL—P. Eles, K. Kuchcinski and Z. Peng
10. Xilinx reference guide, Spartan-3 Libraries Guide for HDL DesignsUG607 (v 12.4) December 14, 2010

MIX
Papier aus verantwortungsvollen Quellen
Paper from responsible sources
FSC® C105338

If you have any concerns about our products,
you can contact us on
ProductSafety@springernature.com

In case Publisher is established outside the EU,
the EU authorized representative is:
Springer Nature Customer Service Center GmbH
Europaplatz 3, 69115 Heidelberg, Germany

Printed by Libri Plureos GmbH
in Hamburg, Germany